TAMPA BAY

Florida in Focus

UNIVERSITY PRESS OF FLORIDA

Florida A&M University, Tallahassee
Florida Atlantic University, Boca Raton
Florida Gulf Coast University, Ft. Myers
Florida International University, Miami
Florida State University, Tallahassee
New College of Florida, Sarasota
University of Central Florida, Orlando
University of Florida, Gainesville
University of North Florida, Jacksonville
University of South Florida, Tampa
University of West Florida, Pensacola

TAMPA BAY

The Story of an Estuary and Its People

Evan P. Bennett

Andrew K. Frank, Series Editor

UNIVERSITY PRESS OF FLORIDA

Gainesville · Tallahassee · Tampa · Boca Raton
Pensacola · Orlando · Miami · Jacksonville · Ft. Myers · Sarasota

29 28 27 26 25 24   6 5 4 3 2 1

Library of Congress Cataloging-in-Publication Data
Names: Bennett, Evan P., author.
Title: Tampa Bay : the story of an estuary and its people / Evan P. Bennett.
Other titles: Florida in focus.
Description: Gainesville : University Press of Florida, 2024. | Series: Florida in focus | Includes
    bibliographical references and index. | Summary: "This book explores the environmental
    history of the largest open water estuary in Florida, revealing how people have interacted with
    nature throughout the long history of Tampa Bay."—Provided by publisher.
Identifiers: LCCN 2023053692 | ISBN 9780813079011 (hardback) | ISBN 9780813070766 (pdf) |
    ISBN 9780813073262 (ebook) | ISBN 9780813080512 (pbk.)
Subjects: LCSH: Estuarine ecology—Florida—Tampa Bay. | Estuaries—Florida—History. | Tampa
    Bay (Fla.)—History. | Tampa (Fla.)—History. | Tampa Bay Watershed (Fla.)—History. | BI-
    SAC: HISTORY / United States / State & Local / South (AL, AR, FL, GA, KY, LA, MS, NC, SC,
    TN, VA, WV) | SCIENCE / Natural History
Classification: LCC F319.T2 B47 2024 | DDC 975.9/65—dc23/eng/20231208
LC record available at https://lccn.loc.gov/2023053692

The University Press of Florida is the scholarly publishing agency for the State University System
of Florida, comprising Florida A&M University, Florida Atlantic University, Florida Gulf Coast
University, Florida International University, Florida State University, New College of Florida, Uni-
versity of Central Florida, University of Florida, University of North Florida, University of South
Florida, and University of West Florida.

University Press of Florida
2046 NE Waldo Road
Suite 2100
Gainesville, FL 32609
http://upress.ufl.edu

To everyone who has worked to save Tampa Bay

CONTENTS

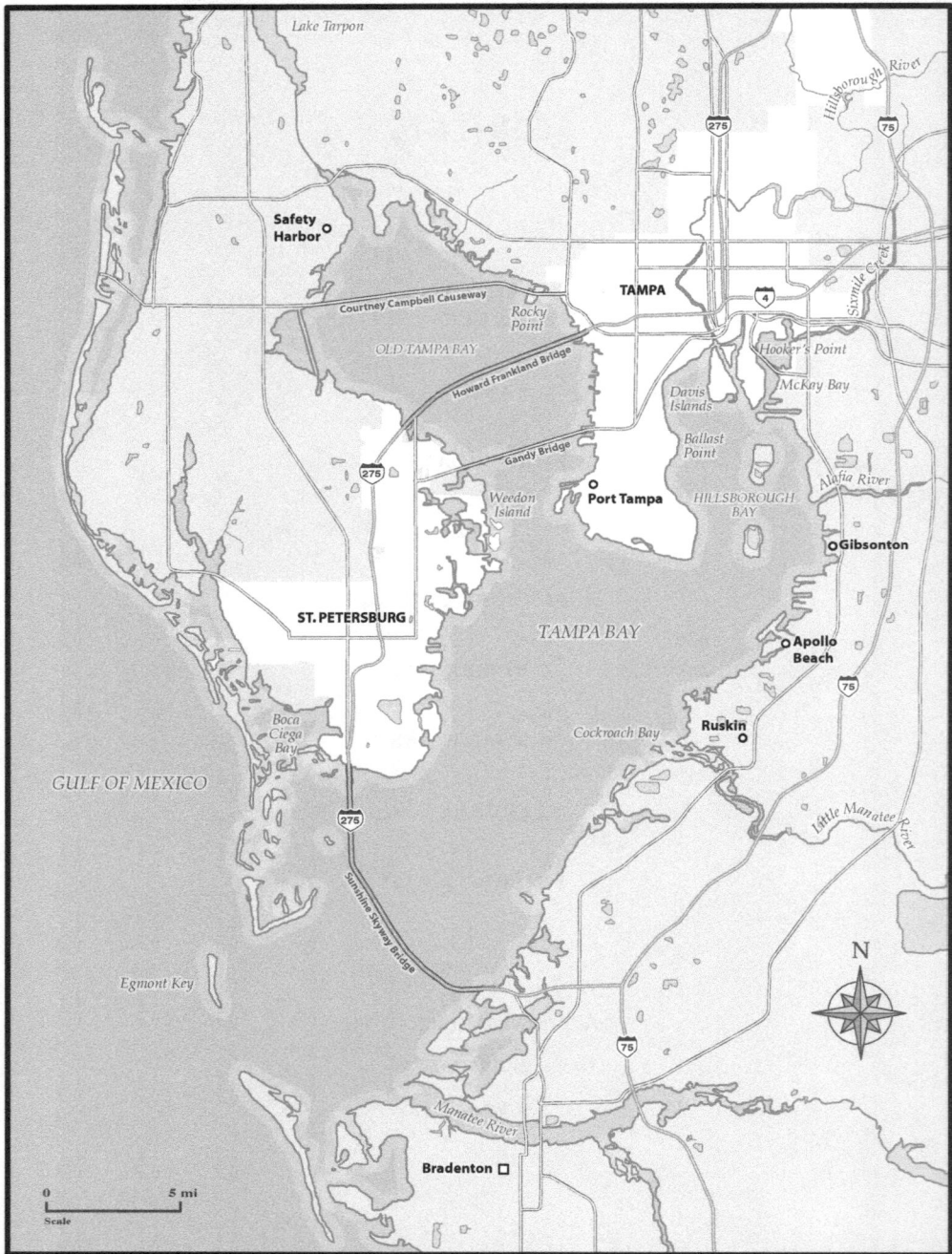

Lake Tarpon

Hillsborough River

75

275

Safety
Harbor

TAMPA

I-4

Sixmile Creek

Courtney Campbell Causeway

Rocky
Point

Hooker's Point

OLD TAMPA BAY

McKay Bay

Howard Frankland Bridge

Davis
Islands

Ballast
Point

Alafia River

Gandy Bridge

275

Weedon
Island

Port Tampa

HILLSBOROUGH
BAY

Gibsonton

ST. PETERSBURG

TAMPA BAY

Apollo
Beach

75

Boca
Ciega
Bay

Cockroach Bay

Ruskin

GULF OF MEXICO

275

Little Manatee River

Sunshine Skyway Bridge

Egmont Key

N

75

Manatee River

Bradenton

0   5 mi
Scale

INTRODUCTION

In late September 2022 everyone in the Tampa Bay Area braced for the worst. Hurricane Ian had passed over the western tip of Cuba as a Category 3 storm and was strengthening as it headed north. Forecasters had Tampa Bay in the dead center of the so-called cone of uncertainty that projected the storm's potential paths. On 26 September, one meteorologist warned of a "potential historic catastrophe" as the storm grew in strength. Expecting they had two days to prepare, authorities issued evacuation orders. In Hillsborough County, 300,000 people were told to leave their homes. Most in southern Pinellas County were told to do the same, as were residents in coastal areas of Pasco and Hernando Counties.[1] People pored over flood maps as television news reporters explained the difference between high waves and storm surge. September 27 brought mixed news: the storm had strengthened, but its path seemed to be turning east. Local authorities warned people not to become complacent. The next day, Ian's sustained winds grew to greater than 150 miles per hour as its path turned further east. Rather than coming up Tampa Bay, the hurricane slammed into Cayo Costa, Sanibel, and Fort Myers, killing 152 people and causing an estimated $113 billion in damages.[2] Thousands across the Bay Area lost power and heavy rains inundated low lying areas, but fears that the storm might push Tampa Bay far inland never materialized. In fact, Ian's outer bands blew the bay in the opposite direction, exposing large areas of bay bottom. Adventurers frolicked on the mud off Bayshore Boulevard, just as they had five years earlier when winds from Hurricane Irma, another storm that appeared to be headed for Tampa Bay before turning east and landing further south, did the same thing.

Two near misses within five years have underscored the connections between water and people in the Tampa Bay Area. But the danger of hurricanes is only one aspect of the relationship. Indeed, ocean storms have figured as only a small part in Tampa Bay's human history. Since the bay's largest recorded hurricane landed in 1848, only two tropical cyclones have landed directly at Tampa Bay, the last one in 1921. The possibility of a storm remains—in fact, becomes more probable with every hurricane season—so asking what Tampa Bay might do to people (and all they have built) is important.

But the environmental history of Tampa Bay is better understood by flipping the question and asking, what have people done to Tampa Bay? Ecologically, Tampa Bay is an estuary—the largest open water estuary in Florida—where seagrass meadows, tidal salt flats, mangrove forests, and waters of varying salinity once supported abundant communities of fish, birds, reptiles, and mammals. But five hundred generations lived in Florida before the bay took its modern shape 2,000 years ago, and people have lived around it ever since. People have always been part of its ecology, have always been participants in its history. Beginning about 150 years ago, their/our place in that story changed radically. As people dredged the bay and dumped chemicals in it, as they hunted some species to near extinction and drove out others, they remade Tampa Bay. The results were often catastrophic. But not all human actions have been negative. In recent decades, people have fought to clean up the bay and strike a balance for nature and people. For bad or good, human fingerprints are all over Tampa Bay.

This book is an attempt to capture those fingerprints. While organized in a roughly chronological way, this is not an exhaustive history of Tampa Bay. Instead, with the understanding that people's relationships with environments are multiple and necessarily overlap, it highlights the ways people have interacted with Tampa Bay. Sometimes the bay was their source of food. Sometimes it was where they played and entertained themselves. At other times it facilitated trade and interaction. As the endpoint of the region's watershed, the bay has constantly received the runoff of human societies. It also has served as a shaper of identity, a touchstone for those who call it home. While there are commonalities in the relationships people had with Tampa Bay across time, the dynamics and meaning were necessarily products of their historical contexts. As a child, for example,

I swam and chased fiddler crabs in the very same spot where Tocobaga children did those very same things 500 years earlier. But our experiences were wholly different. The physical setting had changed with time, of course, but meaning, too, changes as historical contexts change. No one steps into the same bay twice.

While people have always interacted with Tampa Bay in multiple ways at the same time, the societies that lived around Tampa Bay—or, in some cases, claimed the bay from afar—prioritized certain interactions. Over the last 150 years, they, and we, have prioritized multiple things at the same time, sometimes at cross purposes. Because of this, it makes sense to approach the history of Tampa Bay in a way that highlights those priorities. That is the logic of the chapters that follow.

Chapter 1—"Larder"—explores how Native Floridians lived around Tampa Bay before the sixteenth century. In that era, people exploited the bay as a food source. They hunted, fished, and gathered from the bay daily, finding in it the bulk of their caloric intake. With such an ample and ready food source, they had no need for agriculture and, over thousands of years, built complex hierarchical societies.

Chapter 2—"Borderland"—begins in the sixteenth century when those societies faced enormous pressure from invading Spanish *entradas*. Sucked up into the global maelstrom that was the Atlantic World, they succumbed over time to warfare and disease even as the Spanish left Tampa Bay a backwater in their massive empire. By the eighteenth century, new people—Englishmen in search of lumber, Cubans fishing for mullet, Seminoles hunting deer, African-descended peoples escaping slavery, and American enslavers jealous to stop them—eyed the bay, hoping to control it. No one did, though, at least not outright, and the bay was more a meeting place of cultures than a home for anyone. When, in the 1820s, the United States ultimately prevailed in the struggle to claim Florida, they turned Tampa Bay from a meeting place to a point of forced departure for thousands of Seminoles dispossessed of their land.

Chapter 3—"Port"—explores how American settlers made Tampa Bay into a place of trade. Naturally shallow and far from everything, Tampa Bay seemed like a poor candidate to become a bustling port, but by the late nineteenth century, thanks to the arrival of the railroads and the discovery of phosphate east of Tampa, it was on its way. In the twentieth century, massive dredging and wharf-building projects made Tampa one of the busiest ports in the South and the largest phosphate port in the world. Thousands in the area tied their livelihoods to the port, and the demands and desires of shippers shaped the politics of the region.

Chapter 4—"Playground"—moves the focus from the people who saw Tampa Bay as maritime gateway to those who saw it as watery getaway. In the early twentieth century, Americans turned their eyes to the coasts as spaces for recreation. Tampa Bay became a vacation spot for some and a new home for others. But what they made was not exactly paradise. Building castles by the shore required moving more than a few buckets of sand, and subdivisions built from dredged seagrass meadows and mangrove forests remade Tampa Bay's shoreline as dramatically as ship channels altered its bottom. At the same time, access to the shore revealed the inequities of Jim Crow segregation and ethnic biases. As the number of people moving to the Bay Area accelerated in the decades following World War II, environmental and social problems both compounded, further altering the coastline.

Chapter 5—"Sewer"—digs into the ecological consequences of growth for Tampa Bay in the twentieth century. As in port cities everywhere, residents and political leaders alike assumed effluent could be handled best by dumping it in the water. But it became clear as early as the 1910s that Tampa Bay was ill-suited to be a sewer. In Tampa and St. Petersburg alike, providing adequate sewerage became a never-ending struggle. New suburbs and their inadequate facilities only added to the problem. Nitrogen from effluent water was only part of the problem, though. Phosphorus rolling down the Alafia River from phosphate processing plants east of the bay and raining down from the phosphate elevators at its docks mixed with the nitrogen to make a nearly perfect poison for killing Hillsborough Bay. Eutrophication—the removal of oxygen from the water—chased off the fish while turbidity, the product of both massive algal blooms and sediment stirred up while dredging, shaded out the seagrasses. Uncontrolled growth was killing Tampa Bay.

Chapter 6—"Cause"—tells the story of those who hoped to save Tampa Bay from these excesses. Angered by the dredging, tired of the smell, inspired by the nascent environmental movement, and informed by the latest in marine science, people from around the Bay Area found in Tampa Bay a cause around which they could unite. They pushed political leaders to do better, sued to make companies obey the law, and rallied to support policies and people dedicated to cleaning up the bay. They stopped the bleeding and got the bay on life support.

Chapter 7—"Estuary"—extends the story of Tampa Bay to the present day to show how activists, political leaders, and government agencies established a model for bringing Tampa Bay—and by extension other coastal spaces—back from the brink. Articulating a scientific model that allowed people both to understand the bay's ecological complexity and to build a model of cooperation, they established a framework for saving the bay and keeping it clean.

The history told in the following pages is not designed to be exhaustive, either of the history of the bay or of the history of particular events. It is, like all histories, incomplete, idiosyncratic, and interpretive. There is still much to be said about the history of Tampa Bay and I look forward to scholars diving in to help us understand it more completely. Ultimately, I hope this book serves as a starting point for discussions, scholarly and public, about how people live in coastal spaces. I believe the people who have loved and who today love Tampa Bay have created a model of how to clean up and manage a coastal environment in ways that benefit both nature and people. And in a time when we're feeding lettuce to manatees starved of seagrass in the Indian River Lagoon, looking for blue green algae blooms in the St. Lucie and Caloosahatchee Rivers, hearing near-constant red tide warnings on Florida's west coast, figuring out how to get fresh water to Florida Bay, and watching the reefs of the Florida Keys bleach before our eyes, some lessons in how to do it right are in order.

1

LARDER

Before there was a bay, there were people. Lower sea levels during the last Ice Age, when people first moved to the area, exposed more of the Florida Platform, making the peninsula much wider than it is today. At the peak of the last glacial maximum, about 21,500 years ago, the modern Bay Area was roughly 120 miles from the Gulf of Mexico. There was no hint of its modern form. The entire peninsula was cooler and drier. Much of what is covered by water today was then grassy savannas dotted with stands of scrub oaks and pines. Wetlands developed in low-lying areas. When workers dredged the bay near Gadsden Point in 1921, they found a quarter-mile-long swath of ancient cypress trees, some upward of seven feet in diameter. South of there, in what is today the naturally deepest portion of the modern bay, was once a freshwater lake. In 2002, scientists took core samples from Middle Tampa Bay and confirmed that this portion of the bay formed from a sinkhole in the underlying limestone. As the Gulf of Mexico rose over several thousand years, the lake and surrounding region became a saltwater bay.[1]

Fresh water attracted mastodons, saber-toothed cats, and other megafauna. The animals attracted people. The shore of the Gulf of Mexico was likely more than fifty miles away when Paleoindians came to the area that became Tampa Bay 12,000 or more years ago. Collectors have found ancient artifacts around the Bay Area, including at what are now offshore sites. Indeed, most traces of Paeloindians are buried under water and sediment, making it difficult to say much about them over time.[2]

The bay took its modern shape as rising seas and increased rainfall brought salt and fresh waters together in the bay basin. The Gulf of Mexico reached the mouth of Tampa Bay about 5,000 years ago. As a warming earth generated more rain, fresh water accumulated as new creeks and rivers delivered it to the area from an expanding watershed. About 2,000 years ago, the bay took its modern shape and composition. Freshwater wetlands gave way to seagrass beds, mangrove forests, and saltwater marshes. Marine species, especially those that thrived where fresh and salt waters meet, replaced freshwater species, although some, like alligators and turtles, continued to thrive in the margins of salt and fresh water.[3]

The people around these waters made their lives gathering marine resources. At the Culbreath Bayou site on the east side of Old Tampa Bay, archeologists found projectile points and scrapers made of chert, clay pottery, and middens made up of thousands of discarded oyster shells. Evidence suggests people lived there and at countless other spots along the evolving shoreline for centuries.[4]

By about 2,500 years ago people around the bay developed what archaeologists identify as the Manasota culture. Diffused throughout the area from Sarasota Bay to the Anclote River, the Manasota peoples relied on the sea more than the land. Tampa Bay's long coastline and diverse ecological niches offered myriad places for settlement and likely supported a large, dispersed population in this era.

The Yat Kitischee site in Pinellas County, near modern-day Feather Sound, represents a typical small hamlet of the Manasota era. First established about 100 BCE and inhabited continually until about 1200 CE (1,300 years!), the site was likely home to eight or ten households—upward of sixty people—at any one time. They lived in small semicircular structures made of wooden posts and covered with either palm or palmetto fronds. Few things in the water escaped their grasp. Using stop nets likely made of palmetto cordage, they caught nearly everything available in the bay: catfish, trout, kingfish, grouper, mullet, sheepshead, drum, sharks, and rays. Manatees also appeared on their menu. And they ate (literally) tons of oysters. Their tools also came from the sea. Conch shells made good hammers; whelks could be turned into anything from dippers to barbs to cutting tools; bivalve shells could weight fishing nets or scrape wood; olive shells made beautiful beads.[5]

The ever-changing bay forced Manasota peoples to be flexible. A changing shoreline during a period of global sea level fall around 400 CE led people to move their houses closer to the water. Settlers at what is today known as the Bayshore Homes site along Boca Ciega Bay, meanwhile, abandoned the site entirely when sea levels fell; their descendants returned 500 years later when salt water once again flooded the bay.[6]

But change worked both ways. People also altered the shoreline. Sometimes alterations were incidental to everyday life, as in the piling up of shells and bones into middens. In other cases, people made changes intentionally. Most notable was the construction of increasingly large mounds. After 100 CE, Manasota peoples began to inter their dead in mounds made of shell and sand.[7] The mounds represented the growing influence of the Weeden Island cultural complex. Archeologists named this cultural complex for Weedon Island, near the mouth of Old Tampa Bay in Pinellas County, where excavations in the 1920s first revealed distinct artifacts, but its roots were in northwest Florida. (The different spellings are the result of discrepancies in the spelling of Weedon/Weeden at the time.) Increased trade between the two groups explains the presence of Weeden Island artifacts at Weedon Island and elsewhere around Tampa Bay. Over time, the line between Manasota culture and Weeden Island culture around Tampa Bay blurred as the two cultures influenced each other.[8]

The number of mounds grew after 400 CE. Ritual rather than practical alterations of the natural world, they were built for two purposes, either as places of burial or as platforms for the houses of leaders. The mounds were not just piles of sand and shell but could be quite complex and reveal a significant level of care in their building. Over time, they, and the burials associated with them, became more elaborate. This reflected changing social dynamics among the Manasota peoples. The bay's ample food supplies not only fed larger populations but supported the development of hierarchies for controlling resources and ritual for explaining or supporting the hierarchies. The presence of marine artifacts as grave goods demonstrated connections between the bay and people.[9]

By about 900 CE, the Safety Harbor cultural complex supplanted the Manasota around the Bay Area. Named for the town on the west side of Old Tampa Bay where archeologists first described it, the Safety Harbor cultural complex did not represent a shared ethnic identity or a common society. Instead, it was a broad range of shared cultural practices, including technologies (like pottery), social organization, and understandings of the natural and spiritual worlds.

Influenced by the Manasota, Weeden Island, and additional cultural complexes from Southwest Florida, the Safety Harbor cultural complex developed within the historical and ecological contexts of Tampa Bay.

Figure 1. Indian mound somewhere near Tampa, Florida, c. 1910. Mounds like this one were a common sight around Tampa Bay until the twentieth century, when many were destroyed by road-building and other construction projects. Built of layers of sand and shell, they served ritual purposes as either platforms or places of burial. Note the person at the top of the mound in the photo to gain a sense of the size. Several mounds remain around the Bay Area, including at Madira Bikel Mound State Archaeological Site at Terra Ceia, Indian Mound Park in south St. Petersburg, and Philippe Park, the presumptive location of Tocobaga, in Safety Harbor. Credit: State Archives of Florida, Florida Memory.

Like their ancestors, Safety Harbor peoples ate mostly what they could catch in the bay or hunt or gather on its shores. Unlike their ancestors, later Safety Harbor peoples ate corn, a crop, which along with squash and beans, they received through trade with farmers in North Central Florida. Safety Harbor towns were physically more complex than the hamlets of earlier generations, consisting of a platform mound with a plaza at its base, houses built around the plaza, and a burial mound, often at some remove. The construction and placement of the mounds and plazas ritualized a worldview in which the physical and the spiritual, the natural and the supernatural, life and death were intimately connected. The mounds also demonstrated the existence of complex hierarchies of power. Each town had a head man who was the political leader of the community. He controlled the town's resources, including its labor, and its diplomatic and military affairs.[10]

Tocobaga, located in what is now Philippe Park in northeast Pinellas County, was typical of Safety Harbor towns. Known by the name of the leader of the community in the sixteenth century when the Spanish first described it, the town sat alongside Safety Harbor, the largest of the numerous shallow inlets that make up the northern coastline of Old Tampa Bay. Two points, today called Booth Point (on the east) and Philippe Point (on the west), formed the mouth of the inlet. (A hurricane in 1848 wiped away part of Philippe Point, so the mouth of the inlet was probably narrower then.) The inlet was a productive estuary, likely rimmed by mangrove forests, grass flats, and intertidal saltwater marshes on its shore. Fish, turtles, marine mollusks and invertebrates, and marine mammals found food and shelter in its seagrass beds and mangrove roots, while wading birds and seabirds built their rookeries in the mangrove canopy and on the shore. Fresh water flowed from all sides, balancing the salinity perfectly for the growth of oyster beds. The people of Tocobaga ate oysters incessantly—some as large as six inches in length—and collected their pearls. They also dived for crown conchs, sea snails, and whelks. They used lines, nets, and perhaps spears to catch all kinds of fish, including sharks.[11]

The residents of Tocobaga built their platform mound on the bluffs of Philippe Point. (It is still visible and a popular recreational spot today.) Twenty feet tall and 150 feet across, the mound is made of alternating layers of shell and sand. At the peak is a level plateau of roughly one hundred by fifty feet. A structure made of wooden posts, likely the house of the headman and his family, stood on this plateau. Nature and humans have reshaped the northern and eastern sides of the mound since the days of Tocobaga, but the ramp on the west side is still apparent. Today covered by trees, the mound once gleamed in the sunlight, a symbol of authority, a visual demonstration of the power to organize the labor necessary to alter nature. Its vistas were commanding, giving an unobstructed view of Old Tampa Bay as well as the surrounding countryside. To anyone who approached by water, the mound was both a landmark and something of a warning.[12]

Tocobaga's houses spread along the shoreline on either side of the mound. The town center is no longer visible, buried under sediment and partly inundated by the waters of Tampa Bay. Centered on a plaza, the town contained houses of thatched palm or palmetto fronds. Spanish accounts record that they were large, communal structures. Beyond the town site, about a quarter mile from the platform mound, was the burial mound. (Archeologists excavated the site in 1930, desecrating more than 1,400 burials in the process.) For more than six hundred years, generation upon generation twisted palm fibers into nets and bowstrings, turned shells, stones, and bones into tools, shaped logs into canoes, and turned clay into pottery, including figures representing the animals they saw in and around the bay.[13]

We don't know exactly how many Safety Harbor towns there were around the bay over time. The destruction of many of the mounds that once circled the bay for roadbuilding materials in the early twentieth century—which literally scattered the remains of the Safety Harbor culture and its predecessors—makes it nearly impossible to re-create the extent of settlements or understand their chronological development. Archaeologists have identified fourteen surviving platform mound sites in the region between the Anclote and Manatee Rivers, but there may have once been as many as eighteen still extant before the era of road construction. Spanish accounts suggest they were similar in design. The town near modern-day Jungle Prada, opposite Johns Pass, had one house capable of holding three hundred people. Uzita, near the Little Manatee River, had a platform mound with the head man's house on top of it, seven or eight large houses in the town center, and a "temple" adorned with a "wooden bird whose eyes were gilded."[14]

Tocobaga was, by all indications, a later Safety Harbor town, first occupied somewhere in the tenth century CE. Beginning around 900, the number of Safety Harbor towns grew in number, spreading around the bay's long coastline. This was a product of both natural population growth and evolving political arrangements that were related to the control of resources. While the bay's bounty seemed limitless, each section of coast functioned as a microenvironment with its own carrying capacity. Too many people in one place could overtax even a very productive ecosystem, so making sure there was enough to go around was a major concern for each group of people along the bay as population pressures grew. Scholars estimate that each town controlled about fifteen miles of coastline and then twenty miles or so of inland terrain, although by the sixteenth century, a few leaders gained control of larger areas.[15]

At its greatest extent in the 1500s, the Safety Harbor culture diffused throughout most of West Central Florida, from the Withlacoochee River to Charlotte Harbor and inland to modern-day Polk, Hardee, and DeSoto Counties. Specific ecological and historical contexts shaped everyday life across this broad region, but the peoples the central Gulf Coast nevertheless shared a broadly similar culture. They perhaps spoke several languages—the people of Uzita and Mocoso spoke different tongues despite their close proximity. They also provided for themselves in different ways—northern Safety Harbor peoples were farmers, as well as hunters and gatherers. Nevertheless, an observer would likely have noticed more similarities than differences among Safety Harbor towns: shared pottery styles; a shared tradition of mound-building; and shared rituals.[16]

Why did the Safety Harbor culture radiate from the shores of Tampa Bay? Part of the answer is no doubt the growing complexity of Safety Harbor politics. The fixed nature of artifacts should never be mistaken for a lack of dynamism among the people who left them. Spanish records convey hints that suggest complex political relationships tied some towns together and created rivalries between others that sometimes led to war. Late Safety Harbor peoples were not so tightly bound to their villages that they could not reach beyond them. At the same time, trade goods traveled even farther than people, connecting the Safety Harbor peoples of Tampa Bay both to other Florida Indians and to regions far away from the Gulf Coast. Goods and ideas flowed both into and out of the Bay Area.[17]

For those who lived along the bay, everyday life moved with the winds and tides, and the migration of mullet down the bay shaped the seasons as powerfully as the movement of the sun from tropic to tropic. Most people no doubt lived their entire lives within earshot of the bay's noisy rookeries. But their world was bigger than that of their ancestors. The waters that surrounded them, both the salt water of the bay and the Gulf and the fresh water of the rivers that fed them, carried goods and people and ideas. For a long time, this worked largely to their benefit. In sixteenth-century Florida, however, death and destruction came by water, and Tampa Bay was one of its chief ports of call. Few survived, and the bay they once knew changed, too.

2

BORDERLAND

Diego de Miruelo, not Juan Ponce de Leon, was the first European known to have landed on the Florida peninsula. Blown northwest while on a mission to enslave Tainos in the Bahamas in the first years of the 1500s, Miruelo landed somewhere on the Atlantic Coast. Reports of his "discovery" soon traveled through the Spanish Caribbean. By the time Florida appeared on maps in 1511—as "Bimini"—enslavers had been capturing Florida Indians along the coast for some time.[1]

Those forays flouted Spanish law, so Ponce sailed to Florida in 1513 with a royal charter to establish Spanish authority. Rather than order, his brief expedition brought chaos for Florida's Native peoples. Spanish captains mounted an unknown number of illicit slaving expeditions into Florida after Ponce's foray.[2] Miruelo was among them. He explored the Gulf Coast in 1516 and was likely the first European to sail into Tampa Bay. Three years later, Alonso Alvarez de Pineda was the first to put the bay on a map. He named it Bahia de Miruelo. Its coordinates were wrong. Spanish sailors found it anyway and began making unrecorded contacts with the peoples of the bay.[3] In 1521, Ponce sailed again to Southwest Florida. Mortally wounded in an engagement with the Calusa, he failed to establish a base of operations. After that, most potential conquistadors and settlers found prospects in Mexico far more enticing than the Florida coast.

The Spanish defeat of the Aztec empire in 1521, and the subsequent flow of treasures to Spain, brought a shipping revolution in the Gulf of Mexico. Prevailing winds and currents required Spanish ships leaving Veracruz to sail north into the Gulf before turning east so they could catch the southerly currents off the Florida coast back to Cuba. Despite the increased traffic in the Gulf, the Spanish still knew little of the Florida coastline. The Gulf Coast offered little to cartographers but trouble. It was low and shallow. It proffered few natural harbors and scant landmarks. Native Floridians resisted the Spanish, leaving them without access to sources of local knowledge. Storms were common. The wreckage of ships lost at sea followed the currents. Flotsam, jetsam, and shipwrecked sailors littered the beaches of the Gulf of Mexico. Florida's coastal Indians collected them and adapted the goods to fit their cultures. The coast was only rarely a refuge for Spanish mariners; Florida Indians usually enslaved or killed them.

Hoping to gain control of the Gulf Coast, King Charles I of Spain awarded Pánfilo de Narváez a land grant that covered its northern rim. Narváez, hoping to settle on the western side, set out from Spain in June 1527 with five ships loaded with a company of six hundred male and female colonists, soldiers, several priests to work among the Spaniards, Franciscan missionaries to convert the Indians, and enslaved Africans to do the hardest work.[4] Narváez hired Diego de Miruelo to pilot the expedition, figuring he knew the Gulf Coast as well as anyone. Sailing from what is now Cienfuegos, Cuba, the expedition plodded for a month along the Cuban coast when a storm blew the fleet into the Gulf of Mexico. Ill-supplied to sail across the Gulf, Narváez sought to land in Florida. Miruelo turned the expedition toward Tampa Bay but failed to find the entrance. The ships instead slipped between the barrier islands and the Pinellas peninsula into Boca Ciega Bay, where they saw a Safety Harbor town near what is today called the Jungle Prada mound.

The people of the town, likely tributaries to Tocobaga, "made many signs and menaces" toward the Spaniards and evacuated. Narváez's soldiers captured four Indians and forced them to lead them to food stores. They took the Spanish to Tocobaga, where they grew more violent. When members of the expedition noticed a collection of Spanish shipping crates, each holding a decaying body of a dead European sailor wrapped in painted deer skins in preparation for future burial, they asked Narváez to burn the boxes, which he did, bodies and all. According to one account, Alvar Nuñez Cabeza de Vaca, the expedition's treasurer, was so angered by the perceived idolatry that he cut off the nose of the town's head man and threw his mother to the expedition's war dogs. When asked about the boxes and some other goods that looked to be from Mexico, the people of Tocobaga explained they had come either from a shipwreck or from the region of the Apalachee in northern Florida. (Spanish sources conflict on what was said; either case was possible.) Wherever they came from, the boxes were harbingers of a changing Bay Area.[5]

Fortunately for the people of Tocobaga, the expedition stayed just two weeks in the Tampa Bay Area. Narváez then led a force in search of food and a port they still believed lay to their north while the rest remained on the ships and sailed northward to meet them. They never rejoined each other. Miruelo eventually gave up looking for the marchers and sailed south for Cuba. Narváez's expedition ended its journey eight years later when four survivors, including Cabeza de Vaca, straggled into Mexico. Narváez died in a storm along the Panhandle coast.

Narváez's sojourn around Tampa Bay was brief, but significant. While he was never sure where he was, landing at Tampa Bay clarified European maritime knowledge of its location. This seemingly mundane accomplishment was an important step in connecting the bay to the larger Spanish empire in the Americas. The Bay Area became a regular, if usually undocumented, stop for ships traversing the eastern Gulf of Mexico. Meanwhile, Cabeza de Vaca's epic account of the expedition tantalized Spanish elites. While he recounted little about the Bay Area, his account led the Spanish to think of the bay as an entrance to Florida, a place where they might build a base of operations in the future.

Hernando de Soto certainly thought the bay would work as a starting point for his 1539 entrada. Given a royal *asiento* to settle Florida in 1537, de Soto spent a year planning the expedition. He assembled a diverse group of 650 to 700 Spanish settlers for the endeavor. Among them were a sawyer, a boatwright, and a carpenter, skilled workers who could turn Florida's natural world into Spanish necessities. There were also at least two women and three men of African descent. Before leaving Cuba, de Soto sent fifty men ahead in a caravel to scout out the bay; they returned with four kidnapped Indians to use as translators and guides.[6] Aware that the people around the bay did not have corn stocks to pilfer, de Soto loaded the ships with hogs.[7] The expedition also brought roughly two hundred horses, an unknown number of herding dogs to catch the hogs, and dogs of war to capture Indians.[8] All carried with them microbes that changed Florida's disease environment.

De Soto's fleet of nine craft left Cuba in May 1539. Despite improved knowledge of the bay's location, the pilots had some difficulty entering the mouth once they anchored off Longboat Key. They knew little about its tricky, narrow entrances, and the low, flat coast offered few obvious landmarks. Even the sailors who had recently reconnoitered the bay had a difficult time figuring out where there was water enough to sail their deep-draft ships. After some searching, they finally found what is today the Southwest Channel, south of Egmont Key. After five days of scraping bottom, the entire fleet made it into Tampa Bay. Once in, they faced another challenge: the bay's coastline. Mangrove forests and salt marshes spread across Terra Ceia Bay on their right while the expansive bay revealed no land at all on their left. They finally found an opening on a sandy beach at Piney Point, south of the Little Manatee River, but the shallow grass flats made it impossible to get the large ships close to the beach. Unloading took days. Once landed, de Soto renamed the bay Puerto del Espírito Santo to honor the holy day on which they landed. Mapmakers repeated this name for nearly three centuries.[9]

De Soto pursued a violent conquest. His soldiers scoured the shoreline for several miles in search of Indians to enslave before the ships were even unloaded. They killed two Indian men who resisted their attempt to enslave them. Having established a foothold, the expedition moved north and invaded the town of Uzita. Warned by signal fires, the inhabitants abandoned the town before their arrival. De Soto's soldiers ransacked it, taking whatever they could find of value. They tore down the temple and the seven or eight large houses and used the materials to build smaller houses for the members of the expedition. De Soto also ordered all the trees around the town chopped down to deny the Uzitans any cover should they approach.[10]

De Soto's men continued to enslave Indians after establishing the expedition's base camp. They traveled on horse and foot around the bay's shoreline and likely used small boats to traverse the bay itself. When de Soto received a report that the people of Uzita were sheltering at another town, he rushed his forces to capture them. Finding the people had fled, he ordered that town burned and loosed the dogs on the Indian guide who had led them there.[11] Uzitans and their neighbors hid in marshes and swamps and used forests for cover while harassing Spanish soldiers with arrows. Their familiarity with the bay gave them ready access to food almost anywhere. The dispersed nature of their towns also worked to some advantage by giving the Spanish few sizable targets to attack, but a lack of broader organization also made it hard for them to organize resistance. De Soto's entourage captured and killed dozens in the roughly forty-five days they spent around Tampa Bay. This was not enough for Vasco Porcallo, one of de Soto's lieutenants, who went back to Cuba when he could not enslave the masses of Indians he had envisioned.[12]

People who had once known only the area around Tampa Bay found themselves caught into a larger Atlantic World as a result of de Soto's invasion. Among them was a young Tocobaga woman the Spanish called Madalena. Enslavers captured her in the summer of 1539 and forced her to march north to Apalachee. When de Soto marched north from there, Madalena sailed to Cuba as a gift for his wife, who then took her to Spain. When de Soto's wife died, Madalena sailed back to Cuba, where she came to the attention of a Dominican friar named Luis Cáncer de Barbastro, who was planning a mission to Tocobaga. Madalena returned to Tocobaga with Cáncer a decade after Spanish enslavers had captured her. Abandoning the Spanish, she returned to her community and her earlier way of life when some men from Tocobaga killed Cáncer and his fellow missionaries shortly after their arrival.[13] Madalena's story was unique, but she was not alone. Other Safety Harbor peoples found themselves living, and dying, in distant places.

Those who remained faced a changed world, too. Some towns, like Uzita, disappeared forever, their people scattered or dead. The leaders of Tocobaga, meanwhile, took advantage of the disruptions to increase their sphere of authority. By the 1560s, they challenged the Calusa to the south for authority on the west coast. The nature of the rivalry or how long it lasted is unknown, but Spanish sources suggest it was bitter. In 1565, the Spanish Crown charged Pedro Menéndez de Avilés to take control of Florida. Menéndez, who established the capital of the Florida colony at St. Augustine, attempted to extend Spanish authority by positioning soldiers in key Indian towns around the peninsula. This included Mound Key, the center of the Calusa world, where he built an alliance with its head man, whom the Spanish called Carlos, and at Tampa Bay among the people of Tocobaga. In March 1567, Menéndez convinced Carlos to lead him and thirty Spanish soldiers to Tocobaga. Carlos repeatedly asked Menéndez to destroy Tocobaga, or at least allow the twenty men he brought on the mission to do it. Menéndez instead promised to seek the release of a dozen or so Calusa the soldiers of Tocobaga had captured; one of the captives was Carlos's sister.[14]

The first encounter between Menéndez and the leader of Tocobaga reveals the extent to which Tampa Bay had become part of a Spanish-Native borderland in the sixteenth century. Menéndez, having arrived unseen at

Tocobaga in the night, arranged for a Spanish captive of the Calusa, who could speak the language of the people of Tocobaga, to approach the town by boat at dawn and wake the sleeping town with greetings of peace delivered "in a loud voice." Startled by the shouting and the sight of six brigantines in Old Tampa Bay, the townspeople fled. The head man remained, however, and sent an enslaved Portuguese sailor to greet Menéndez. The sailor explained he had been shipwrecked six years earlier when his small boat carrying blankets, corn, and honey from Campeche to Veracruz had somehow been blown across the Gulf of Mexico. The people of Tocobaga had killed his shipmates, but he had been able to hide, surviving on "palm hearts, acorns, and some shellfish" for a month. Indian fishers found him and delivered him to Tocobaga. Through the Portuguese sailor, the head man explained that he would "sooner die" than abandon "his house of his prayer and his gods," making clear he was familiar enough with Spanish treatment of Native people in the past.[15]

But the Spanish approach to conquest had changed since the times of Narváez and de Soto. Menéndez hewed closely to the requirements of the New Laws (1542), which prioritized conversion and alliance and placed severe restrictions on enslaving Native peoples. In his mind, Native Floridians should accept Christianity and become vassals of the Spanish monarch. Even when the head man of Tocobaga called together more than 1,500 armed men from settlements around the Bay Area for the negotiations, Menéndez assumed he was setting terms.[16]

The people of Tocobaga did not resist Menéndez openly but instead agreed to allow thirty Spanish soldiers to remain among them. Understanding that the Calusa had allied with the Spanish, they knew that if they attacked Menéndez's forces, the Spanish would retaliate, and Carlos would gain the upper hand in their ongoing struggle for control of the west coast. However, if they agreed to ally with the Spanish, the Spanish would leave them alone and Menéndez, believing he had settled a dispute between two competing vassals, would place a check on Carlos's designs. It was a way of buying time.

Not long after Menéndez left the Bay Area in March 1567, the people of Tocobaga resumed their rivalry with the Calusa, who were themselves suffering a period of infighting fueled by the Spanish presence. The following January, Spanish officials went to Tocobaga to compel the head man to apologize for "carrying off some [Calusa] towns."[17] When the officials arrived, they found that the people of Tocobaga had killed the soldiers Menéndez left with them. Rather than feeding themselves from the bay and the surrounding countryside when rain ruined their supplies, the soldiers demanded that the people of Tocobaga supply them. Worse, they, in the words of one Native witness, "were very mean-spirited" and "put their hands on everyone."[18] What this meant exactly is a matter of conjecture, but clearly the soldiers assaulted both men and women. To avenge the killings, the Spanish burned Tocobaga and abandoned their attempt to colonize Tampa Bay. Unpressured by direct Spanish contact, Tocobaga rebounded and grew in influence along the Gulf Coast in the late sixteenth century. Apparently allying with the Pojoy, who lived on the south side of Tampa Bay near its mouth (first identified as a distinctive group by the Spanish in the early seventeenth century), they continued to use their canoes to fish and to trade with coastal Indians as far north as St. Marks.[19]

Their attempts to keep the Spanish at arm's length, however, became more difficult as Franciscan missionaries pushed westward from St. Augustine in the early seventeenth century. The Franciscan mission near the mouth of the Suwannee River became a point of contact and tension. In 1611, a group of Tocobaga and Pojoy Indians killed seventeen Native Christians who were delivering supplies for the mission. Spanish officers responded by leading their Timucua allies from North Florida down the coast to Tampa Bay, where they seized the attackers and beheaded the Pojoy's leader. Within a few years, animosity between the people of Tocobaga and the Spanish had subsided sufficiently that Tocobaga boatmen began carrying freight both along the coast and up the rivers of North Florida. By the end of the seventeenth century, people from Tocobaga established at least one town of about 350 people along the Wacissa River, southeast of modern-day Tallahassee.[20]

Why would the people of Tocobaga, who had more or less successfully kept their distance from the Spanish in the nearly half century following the destruction of the garrison in 1568, decide to work for the Spanish beginning in the 1620s? One answer is a desire to benefit from trade with the Europeans. Part and parcel of Spanish colonialism was gift-giving. Florida's Spanish governors maintained good relations with their Indian allies by providing them European metal goods, textiles, and decorative objects. The Spanish would also trade those items for salvaged gold or natural products they could not obtain elsewhere. Archaeological sites around Tampa Bay are awash in the remnants of Spanish trade goods. By the seventeenth century, the bay was a stop for Spanish mariners who wanted to trade for ambergris, the pungent, valuable whale secretion used in perfumes that Indians collected as it washed up on shore.[21] The people of Tocobaga and their neighbors likely figured they could obtain trade goods while maintaining some distance and autonomy by carrying freight for the Spanish.

But is just as likely that the leaders of Tocobaga allied with the Spanish to seek security in a rapidly changing world. The seventeenth century was a period of rapid weakening for Native Floridians. Florida's European population grew in the seventeenth century, and with it the number of potential hosts who could introduce diseases like typhoid, whooping cough, tuberculosis, and smallpox. Most Europeans lived in St. Augustine, but trade routes connected the town to all of Florida. The mission system and the expansion of cattle ranching in North Central Florida provided other sustained points of contact from which disease might spread. As disease spread and struck people of all ages, societies faltered. Old enemies rushed to increase their power or settle old scores. Warfare begot more disease. Survivors looked for security where they could find it.

We have no records to specify how disease traveled in the Bay Area, but the region's population declined across the seventeenth century. The remaining towns around the bay collapsed. At some point in the early seventeenth century, the Pojoy switched their allegiance, either willingly or under duress, to the Calusa.[22] What did this mean for Tocobaga? It is possible that the Pojoy's transformation from ally to enemy meant that the presence of Tocobagans along the Wacissa River was the result of relocation rather than expansion.[23] A 1678 Spanish report gives a picture of a fearful people, living on a concealed branch of the river and afraid to go out to sea, "because they are concerned about their lives and the injuries that they know [their enemies] would do to all of them."[24] The Spanish governor, recognizing their fear and their value as boatmen, assured them they could remain where they were. Notably, they remained resistant to baptism.

As late as 1699, Spanish soldiers noted the continued presence of villages around Tampa Bay. Unfortunately, they did not visit them, so we know nothing about them but their names.[25] In time, the people of those towns faced much the same fate as the people of Tocobaga. Some found protection by moving closer to St. Augustine, but for most, generations of warfare, disease, and displacement spelled an end to their societies. Raiding that accompanied an invasion of militia from the newly formed Georgia colony in 1739 is believed to have brought about the destruction of the Pojoy and an end to the last Native settlements near Tampa Bay.[26]

The desolation of the peoples of Tampa Bay necessarily altered the bay itself. People had always lived by and shaped the bay; their disappearance would have been noticeable. Most changes would have been subtle, but in their totality, significant. Along the shoreline, nature reclaimed Native townsites. Humidity sped the decay of buildings. Fast-growing trees overtook streets and home sites and covered what parts of the mounds the rains didn't erode. Mangrove forests and salt marshes claimed man-made beaches. In the water, fish traps sank beneath the sand. Populations of species that the peoples of the bay had fished, hunted, and collected for millennia expanded without those pressures. Shellfish populations likely grew. Oyster bars grew in size.

Little documentation exists about the bay in the eighteenth and early nineteenth centuries. What we have comes from European and Anglo-American sources, so our view of events from this era is doubly partial. The records of official expeditions provide a slim documentary record and give us fleeting glimpses of human activities around the bay. In recent decades, archeologists have revealed more about those activities, but even so, we are able to chart only the broadest outlines of the bay's history in those years. Nevertheless, the bay had a history. Bereft of the civilizations whose generations had lived along its shores, the bay was not left derelict. Its position at the edges of four empires—Spanish, British, Creek/Seminole, and, later, American—made it a borderland where all four met—sometimes peacefully, often violently—over time.

New peoples invaded the bay in the eighteenth century to take advantage of its resources. Their lives were different from the lives of their predecessors. Until the early nineteenth century, most lived there only seasonally. Their conceptions of the bay, too, differed from those of earlier inhabitants. The new people fished the bay and hunted along its shores, but it was not their home, at least not at first. To them, the bay was frontier space where nature presented exploitable resources that could be taken and used or sold elsewhere. It was also a space that had to be secured from competitors.

The Spanish government sponsored two expeditions to Tampa Bay in the 1750s. British reconnaissance missions around the Gulf of Mexico, including a 1745 stop at Tampa Bay, alarmed Spanish authorities and spurred renewed interest. Juan Bautista Franco arrived in 1756 at the tail end of a longer mission to survey the Gulf Coast's timber resources. He sailed around the bay for twenty-two days in November and December, documenting the size and positions of Old Tampa Bay and Hillsborough Bay, as well as the presence of the Hillsborough and Palm Rivers. The banks of the rivers, he told his superiors, were "covered with . . . sabine [likely pine], live-oak, walnut, and laurel" as well as with "pastures and several watery lagoons, abundant in animals," including "deer, rabbits, bears, and turkeys." The "beautiful and vast plains covered with pine suitable for all types of ship timber" impressed Franco even more, and he deemed that the samples he cut compared to the finest he had seen from anywhere around the Atlantic World. The bay, too, seemed a perfect place to cut timber: excellent stands could be found "no more than two miles" from any river, while the sandy, mud-free banks of the shoreline made it easy to load the lumber on ships. "And what makes it even more desirable," he further explained, "is my not having found a single mosquito in the area through which I went." He wasn't the last visitor lulled into a false sense of security by Florida's winter charms.[27]

Franco's brief account included as many Native people as mosquitos. What he missed, or simply failed to note, was that indigenous hunters were at work on the land around the bay. Francisco Celi, who charted the bay several months later, in April and May 1757, recorded his encounters with Indians. Celi was not surprised by their presence and the Indians were similarly nonplussed by the Spaniards. On several occasions, the hunters rowed their canoes to Celi's ship, and the captain received them with gifts of honey, rum, and tobacco. The Indians carried "their necessary equipment for deer hunting," Celi noted, which included "guns, ammunition, and four hollow deer skulls" that they wore as decoys. The skulls still had their antlers and "some small cords inside which made the ears move about." In their camp, Celi saw "three piles of deer skins laid out and another two piles of venison."[28] The group was small, made up only of men and boys.

While Celi described the hunters as "natives of this country," the Indians he encountered did not live near the bay. They were sojourners. By the 1750s, several Native groups originally from north of Florida vied for space on the peninsula. They hoped to better position themselves between the Spanish in St. Augustine and Pensacola and British traders in Georgia, who sought to draw them into their respective orbits. Yamasee Indians arrived from what is now South Carolina beginning in the 1710s; several

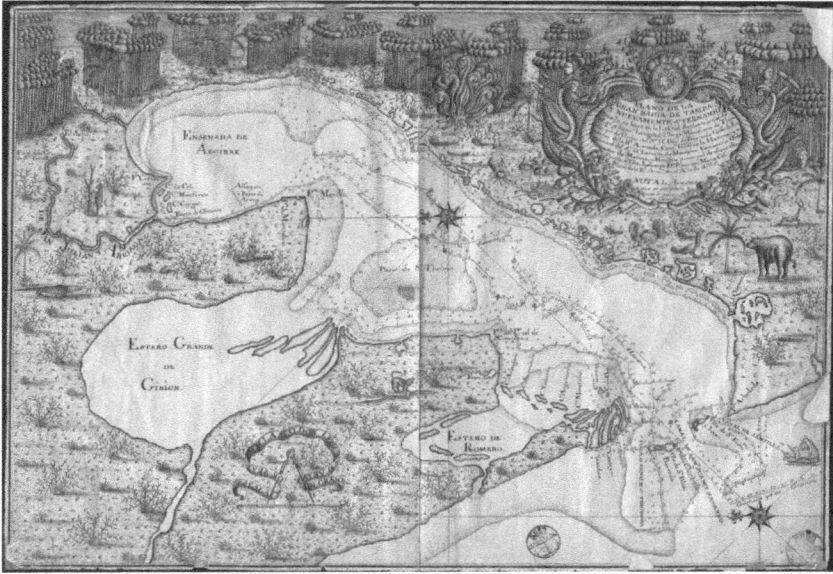

Figure 2. *Plano de la Gran Bahía de Tampa* by Franciso Celi, 1757. There are two versions of Francisco Celi's 1757 chart of Tampa Bay. This one is considered a copy of the first. Celi's charts are the earliest detailed maps of the region. Oriented with east at the top, it details the bay's two lobes, identifies geographic features by names that have been lost to time, and records detailed depth soundings. Its depiction of the eastern shore of Hillsborough Bay is the most detailed part of the map and includes illustrations of the Seminole hunting camp Celi encountered. Credit: Archivo del Museo Naval de Madrid.

groups of Creek Indians migrated from what are now Georgia and Alabama in the decades after. They built towns in the region formerly occupied by the Apalachee, along the Suwanee River, and, by the 1750s, in the Alachua region. By the 1760s, the British identified the Alachua Creeks under Ahaya, or Cowkeeper, as the Seminoles. Other Creek towns developed this identity over time as they separated themselves from the larger Creek sphere to the north. Depending on their preference, Seminole leaders traded with either the Spanish or British, exchanging deerskins for European trade goods, including firearms. The trade had already reduced white-tailed deer populations in many parts of the Southeast, so Florida's sizeable deer populations gave them an advantage. The Bay Area must have been an attractive area in which hunt. Seminole hunters steered their canoes around the bay, built camps along its shoreline, fished from its waters, and hunted its birds, but they were in many ways like the Europeans who documented their presence: newcomers grasping for what resources the bay might provide to distant markets.[29]

The Spanish never sent loggers to tap the Bay Area's forests as Franco and Celi hoped. Instead, Spain's monarch signed over Florida to the British as part of the settlement of the Seven Years' War in 1763. Two years later, the Royal Navy commissioned a survey of Tampa Bay by George Gauld, who was to determine the bay's potential as a port. Arriving in June, Gauld took careful measurements of the bay. He noted an "abundance of fish, oysters, and clams, as well as large and small water-fowls, turkeys, deer, &c. with plenty of fresh water and wood."[30] His assistant, James Cook, meanwhile reported that they saw unoccupied "Indian hutts."[31] These were empty because of the seasonality of the hunt: deer hunting took place in the fall and winter, and skins traded in Pensacola or Charleston in the spring.[32]

On Mullet Key, Gauld caught sight of another group of itinerants who came to the bay to take its resources to distant markets: Cuban fishers. Havana's markets demanded fish and a wide range of maritime animal products. Florida's shallow Gulf Coast estuaries offered about anything they needed. Florida's coastal Indians, especially the Calusa, carried those items to Cuba in their long canoes throughout the seventeenth century, but as their numbers declined, Cuban fishers pressed northward into the Florida Keys and the estuaries of southern Florida. The transfer of Florida to the British did nothing to slow their advance. The fishers—at first, they were all men—established small outposts along the Gulf Coast called *ranchos*. Seasonal camps, they consisted of a few palm-thatched buildings that could be easily repaired at the start of a new season and racks for salting and drying the fish. The fishers loaded their smacks with barrels of salt in August, headed north to Florida, and returned to Cuba with loads of salted fish and roe in March. Neither Franco nor Celi had noted their presence at Tampa Bay in 1756–57, so it is possible they had only reached that far north in the years before Gauld's expedition.

The fishers' sojourn coincided with the busiest time in Florida's estuaries: the striped mullet spawning season, when massive schools of adult fish rush out of the estuaries into the open ocean to spawn. The fishers sought the bony fish mostly for its roe, but the salted fish was profitable, too. More popular were the abundant jacks, redfish, snappers, permit, and pompano that brought high prices in Havana. The fishers also hunted turtles for meat and sharks for oil. It was profitable business: in 1772, a British surveyor reported that Cuban fishers told him that the six schooners employed in transporting the fish to Havana each earned an average of $2,000 per trip and sometimes made two trips a year.[33]

In the second half of the eighteenth century, Spain and Britain twice traded Florida across negotiating tables in Europe. Native hunters and Cuban fishers, however, possessed the Gulf Coast, including Tampa Bay. Over time, the numbers of each group increased, and they developed closer ties to one another. Some Cuban fishers moved to Florida more or less permanently, making the ranchos into year-round settlements. At the same time, Seminoles established towns south of their initial settlements in northern Florida. Their interactions grew more intimate. They exchanged goods, labor, and even vows of matrimony. By the early nineteenth century, the children of those bonds, called Spanish Indians by English-speaking observers, were the backbone of the rancho communities. Archeologist John E. Worth has argued that they "formed their own unique culture" that drew on "different cultural traditions with roots in Spain, the Canary Islands, Cuba, Mexico, the southeastern United States, and perhaps other areas (including Africa)." Theirs was not merely a "hybrid of Spanish and Indian," but a "new culture, a new identity" forged by the natural, social, and cultural environments of the Gulf Coast. Ships sailed between Havana and Bahia de Tampa and the Costa de Tampa with regularity. Indian children born at what was identified as the Pueblo de Tampa were baptized in Cuba. Tampa Bay had become a middle ground where peoples of diverse backgrounds built a unique society out of the pieces of each.[34]

American expansion brought more new people to the Bay Area in the early nineteenth century: escaped African American captives. After 1790, American enslavers pushed further west and south into Georgia and Alabama to take advantage of the boom times in cotton production unleashed by the invention of the cotton gin. This movement brought enslaved people closer to Florida and the potential freedom it offered. Escapees had long made their way to Florida. After the American Revolution, increasing numbers found refuge among the Seminoles. They typically lived adjacent to Seminole towns, where they grew corn and other crops for the towns and helped to herd cattle. Over time, marriages and children bound Native Seminoles and Black Seminoles together.[35]

Beginning in 1812, American leaders waged a campaign of intermittent but vicious warfare that scattered Native and Black Seminoles across the peninsula. American troops crossed the St. Mary's River, captured Fernandina, and then attacked St. Augustine and settlements along the St. Johns. The Spanish held the capitol and retaliated by enlisting the Seminoles to attack plantations owned by American sympathizers. In response, American troops attacked Seminole towns and reenslaved any person of African descent they might find.[36] In 1816, American troops destroyed a force of more than three hundred Black soldiers and Native allies left by the British at the so-called Negro Fort at Prospect Bluff on the Apalachicola River. In 1818, an invading force led by Andrew Jackson attacked Seminole towns near modern-day Tallahassee and along the Suwannee River before capturing St. Marks and Pensacola.[37] As in the earlier invasion, the Americans reenslaved any Black people they captured.

Some who escaped sought sanctuary around Tampa Bay. "It is the last rallying spot of the disaffected negroes and Indians," Captain James Gadsden, Jackson's aide-de-camp, told him in 1818.[38] Cuban fishers invited some into their ranchos. Others established towns around the Bay Area. In 1821, Seminole leader Neamathla recorded at least five Native towns in the vicinity of Tampa Bay. Not large, the towns were diverse. Several, including Tatstalajoatska, or Watermelon Town, and Red Town had been established by Creek Indians ejected from Alabama in the wake of the Creek War of 1813–14.[39] Gadsden, meanwhile, advised that others were Miccosukee and Seminole from northern Florida.

Angola, a community of formerly enslaved African-descended people, sat near the confluence of the Braden and Manatee Rivers, a few miles inland from the mouth of Tampa Bay. Black Seminoles made their way there as early as 1812. Over the next several years, the population grew. In 1815, eighty or so Black volunteers recruited by the British to fight at the Battle of New Orleans arrived, dropped off by the same British commanders who had established the Negro Fort. When the Americans destroyed that fort in 1816, some of the survivors made their way south to the Bay Area. More escapees arrived in 1817 and 1818, propelled by Jackson's invasion. Many of those were families whose escape was secured by the heroic stand of Black Seminole troops at the Battle of the Suwanee.[40] By 1821, upward of six hundred people lived at Angola, where they, in the words of one scholar, "simply tried to subsist by farming and fishing, but strategically defended themselves when required."[41]

In the tumult of the 1810s, Tampa Bay was both a sanctuary and a place to meet allies. Fishers with connections in Cuba brought supplies and transported leaders to Havana. At the same time, British subjects hoping to recruit their onetime Black and Native allies to destabilize American frontier claims came from the Bahamas. In the end, the Seminoles received no official support, but those in Angola benefited from the connections when the Spanish ceded Florida to the United States in 1821.

American enslavers could not abide the presence of a large community of free Black people surrounded by Native allies at the southern fringe of their expanding empire. Despite the protections afforded Florida residents in the treaty with Spain, "some men of influence and fortune, residing somewhere in the western country" (likely a reference to Andrew Jackson and his associates) hired a force of more than two hundred Coweta Creek Indians to "proceed along the western coast of East Florida, southerly, and there take, in the name of the United States, and make prisoners of all the men of colour, including women and children" within months of the transfer of Florida to the United States. The sponsors of the kidnapping raid, an observer noted, "thought of making a speculation in order to obtain Slaves for a trifle."[42] The attacks were devastating. "[T]he terror . . . spread along the West Coast of East Florida," one observer noted, breaking up "all the establishments of both blacks and Indians." In a surprise attack on Angola, the Coweta Creek warriors captured about three hundred people and burned the town to the ground. There is no telling how many residents died in the attack, but hundreds of survivors ran for their lives, some as far south as Cape Sable and Tavernier Key, where Bahamian boats patrolling for shipwrecks rescued them. The Coweta Creek warriors further attacked neighboring Seminole towns and the ranchos along the coast in their search for Black people to sell to American enslavers.[43]

In the minds of American political and military leaders, Tampa Bay could not remain a fluid borderland. The United States government, understanding how the bay and other harbors functioned as connecting points to Cuba and the Bahamas for the Seminoles and their Black allies followed the attacks by banning Seminole settlements anywhere within twenty-five miles of the coast. In 1824, the army established Fort Brooke at the mouth of the Hillsborough River to control the region. In 1831, the territorial council passed a sweeping law that required licenses to sell catch fish intended for foreign markets, imposed enormous fees to register any fishing boat not wholly owned by U.S. citizens, and barred anyone from employing "any . . . Indian, as a fisherman or mariner, or in curing said fish on board said vessel, or on shore."[44]

The number of Cuban-operated ranchos declined rapidly. Filling the void, American William Bunce established a homestead on the site of an existing rancho at Shaw's Point, at the mouth of the Manatee River in 1834. In short order, it became a village with a store, blacksmith and carpentry shops, and more than thirty palm-thatched huts in which his workers and their families lived. Despite the law, he employed a Native workforce, most likely a mix of Spanish Indians and Seminoles.[45]

Bunce's arrival came at a critical moment in American-Seminole relations. In 1830, the federal government passed the Indian Removal Act, which required that Native peoples east of the Mississippi River move to reservations west of it. The U.S. Army began to increase pressure on the Seminoles to leave Florida. In late 1835, war broke out as the Seminoles resisted increased American pressure. The ranchos became a site of conflict. The army, suspecting Bunce was not only employing Seminoles but also encouraging them to stand up to the Americans, destroyed his original rancho in 1837. When he removed to Cabbage Key, the army burned the new rancho to the ground and removed his entire workforce (including several Spanish citizens) to New Orleans for transport to the western reservations.[46] Fishing camps, like the one begun by Antonio Maximo Hernandez at the southern tip of the Pinellas peninsula, returned only after the fighting paused in 1842. A massive hurricane in 1848 destroyed them.

Tampa Bay figured largely in the Second Seminole War (1835–1842) as the site of dispossession for thousands of Seminoles. Fort Brooke doubled as a prison for Seminoles who surrendered or were captured in battle. There they waited for the steamships that would carry them first to New Orleans from whence they traveled via riverboat toward the forced march to what is now Oklahoma. The journey took more than a month. Many did not survive. According to one nineteenth-century account, the army banished 4,420 Seminoles and maroons via Tampa Bay. One flotilla carried more than 1,100 at once.[47] In many cases, the army forced Black Seminoles into the slave markets in New Orleans; slave traders even came to Fort Brooke to enslave them.[48]

For those exiles, the foot of Fort Brooke's wharf at the mouth of the Hillsborough River was the last dry land they touched in Florida. The fading outline of the Egmont Key and Mullet Key was their last sight. "[T]he women," one army officer recorded, "prostrated themselves and kissed the earth, muttering imprecations upon the heads of those who were . . . tearing them from it." Asked by the same officer about his thoughts as he stared into the distance as the boat left the bay, Coacoochee reportedly explained, "I am looking at the last pine tree on my land. . . . Florida . . . was my home. I loved it; and to leave it now, is like burying my wife and child."[49] When the Americans attacked the Seminoles again in 1855, sparking the Third Seminole War, Tampa Bay again became the site of displacement. By that time, however, the United States had built a lighthouse and jail on Egmont Key. In 1858, Billy Bowlegs (Halpatter Micco) boarded the *Grey Cloud* there, leading the last of the exiled Seminoles west.

Seminole removal closed a period in Tampa Bay's history. No longer was it a borderland. The United States, by removing the Seminoles and their allies, ended the contests for power that had shaped the region for nearly three hundred years. It was not quite a domesticated space. That followed in the next century, as Americans worked to shape the bay to their ends.

3

PORT

Beauty and bounty. That's what people saw when they looked at Tampa Bay in the early nineteenth century. To an army officer assigned to Fort Brooke in 1832, the bay was the "most delightful spot" with "an abundance of . . . fish . . . the best of oysters, the large green turtle in great plenty, and water fowl of every description."[1] Englishman George Ballentine, arrived two decades later, agreed. The bay was "beautiful" and "picturesque," with "inexhaustible quantities" of oysters.[2] What people did not see when they looked at Tampa Bay, however, was potential for a great port. Its tricky entrance, its flat shoreline, and, most of all, its shallow depth, seemed to place natural limits on the bay's promise as a deep-water harbor. Ballentine, whose schooner had to anchor six miles from Fort Brooke "on account of the extreme shoalness [*sic*] of the bay," believed the region did "not seem destined to rise very rapidly in importance" because the bay presented a "difficult . . . approach for shipping."[3]

Despite the doubters, Tampa Bay has become one of Florida's, and the nation's, most important ports. By the mid-twentieth century, the docks around the bay handled more cargo than any other in the state and trailed only New Orleans, Baton Rouge, and Mobile among southeastern ports.[4] Ships moved people and freight to and from all parts of the world. This was no accident. Fishers, cattle barons, loggers, railroad operators, miners, longshoremen, and city-builders made the bay into a bustling port by conquering nature in two ways. First, they exploited the surrounding countryside and turned its natural products into commodities. Then, with the help of the federal government, they overcame the limits of its shallow depth by dredging the bay to make it easier to transport those products out to the larger world. To make Tampa Bay a port, they changed the bay forever. The consequences echoed for generations.

Seafood was Tampa Bay's first great export, and it remained so long after the transfer of Florida to the United States in 1821. At Spanishtown, a small settlement on a creek of the same name just west of the mouth of the Hillsborough River, Cuban fishermen maintained a rancho into the 1830s. Antonio Maximo Hernandez, who eventually established a rancho at the southern point of the Pinellas peninsula, arrived before the American takeover and lived full-time around the bay and survived through a combination of fishing and farming. In the 1820s, Connecticut fishermen joined the Cubans, arriving in the wintertime to fish for mackerel, grouper, drum, and other popular species they could sell from their live wells in Havana. Both Cubans and Yankees took stone crabs, oysters, and green turtles and their eggs in massive quantities as well.[5]

The end of the Second Seminole War in 1842 brought a period of population growth and increased maritime traffic, much of it boats and schooners transporting natural products like lumber from the bay's hinterland to larger markets.[6] The Bay Area never developed into a plantation belt devoted to staple crops. A half-dozen or so wealthy planters forced enslaved Black laborers to plant sugar along the Manatee River, but by the late 1850s, every attempt to make sugar plantations near Tampa Bay had failed.[7] Most white settlers in the years before the Civil War were farmers of smaller ambitions. The Bay Area's sandy soils lent themselves best to small-scale agriculture, so slaveholders and non-slaveholders alike instead pursued what might be best called mixed farming. Odet Philippe famously established his small St. Helena plantation at the site of the main Tocobaga town at the north end of Old Tampa Bay with his daughters and a few enslaved people around 1842. They planted a little cotton, established the region's first citrus grove, and grew fruits and vegetables to supply the troops at Fort Brooke.[8]

Most white settlers instead chose to herd cattle. The ecosystems around Tampa Bay proved an ideal environment for the feral offspring of Spanish cattle introduced in the seventeenth century. Francisco Celi noted their presence on his 1757 map but, unfortunately, said nothing about them. Seminoles found security in claiming cattle beginning in the eighteenth century, and control over the vast herds in part fueled the conflict with white settlers in the nineteenth. The dispossession of so many Seminoles after 1835 left the wild cattle stocks wide open to American settlers. More aptly called cow hunting than cattle ranching, the work was difficult. Cow hunters chased the cattle across flatwoods, prairies, hammocks, and swamps. Having caught them, they tangled with their large horns and ornery temperaments to mark them as property. The cattle again foraged on the open range until the round up, at which point the cow hunters pursued them all over again to drive them to market.

Florida scrub cattle were scrawny. It took three to make one shadow, one observer joked.[9] They could, however, turn a profit for those who could organize the capital necessary to round up and transport them from Tampa Bay to Key West, Cuba, and the Bahamas. William Hooker did just this. He owned nearly 1,500 acres east of Tampa where his enslaved labor force of at least nine raised more than 2,500 cattle.[10] From the 1840s until he sold his cattle holdings in 1861, Hooker used the spit of land at the north end of Hillsborough Bay that still bears his name as the loading point for his cattle shipments.[11] Others loaded cattle at Ballast Point. Merchant and ship owner James McKay purchased cattle from around the Bay Area and then shipped them to Cuba. An enslaved crew manned his state-of-the-art, 150-foot, screw-driven steamship on runs to Havana. By 1860, they transported roughly four hundred cattle per month.[12] The more than 7,200 cattle shipped from Tampa Bay in 1859 brought in nearly $100,000 and accounted for almost 60 percent of value of products shipped.[13] The port also handled roughly 2,900 cow hides, 20 barrels of beef tallow, 200 hogs, 16 bales of deer skins, 400 barrels of beeswax, and more than 60,000 feet of lumber. Produce made up the rest of the shipments.[14] Incoming ships, meanwhile, carried consumer items into Tampa.

By the 1850s, Tampa Bay was a regular, if small, port of call for the routes that tied together major U.S. ports on the Atlantic and Gulf Coasts, as well as ports in the Caribbean. Little was done to improve navigation in the fledgling port. Congress passed the first bill providing for a lighthouse on Egmont Key in 1838, but a lack of funding delayed its construction for a decade. Completed in May 1848, the lighthouse suffered significant damage in a massive hurricane the following September. Light traffic meant there was no rush to fix the it; repairs took another decade.[15] Neither the federal government nor local interests sponsored any dredging projects. To transport anything into the growing village of Tampa, large ships anchored at Ballast Point, and crews lightered goods onto smaller boats that could make it through the shallow waters of upper Hillsborough Bay.

During the Civil War, rebel leaders expected Florida to provide beef, salt, and foodstuffs for their armies. Rebel sympathizers constructed several saltworks on the bay and sought to avoid the blockade by outfitting fast ships. In response, the U.S. Navy's East Gulf Blocking Squadron focused on closing Tampa Bay. In 1863, the crew of the USS *Tahoma* captured or destroyed at least six ships. Most, like the *Crazy Jane*, a schooner seized just off Gadsden Point, carried cotton and turpentine. Crewmen from the *Tahoma* also led a raid to burn the *Scottish Chief*, a steamer James McKay was building on the Hillsborough River.[16] In late 1864, detachments from the USS *Stars and Stripes* and four other steamers destroyed "several large boilers and everything of value connected with" McKay's works near Rocky Point, "without a single casualty."[17]

Despite losing its light to rebel thieves from Tampa early in the war, Egmont Key was a beacon to people who sought to escape rebel-controlled areas. By the summer of 1862, a dozen former escapees sought refuge on the island, where they "engaged in clearing the ground and planting sweet potatoes." Unlike white Unionist refugees, who could find ready transportation to the relative safety of Key West, those freedpeople had to supply themselves from the sandy ground and ample waters for the duration.[18]

In 1866, federal authorities relit the light at Egmont Key, reopening the bay as a port. The end of the war brought revival and expansion of commerce on the bay as weary Americans sought out cheap land and the virtues of subtropical living. Land speculators promised that sojourns in Florida could cure all ills. Thousands moved south to try find their place in the sun. Many went to the northwest part of the state because of the ready availability of rail and steamboat connections, but others sought out more remote spaces to make their mark. Tampa Bay, accessible by boat from the end of the railroad line in Cedar Key, attracted those. Hillsborough County's population nearly doubled in the years following the war, growing to nearly 6,000 people by 1880.[19] Traffic on the bay grew. "Boats, big boats, little boats (much in the majority), long boats, short boats, row boats, sail boats, and perhaps a steamboat or two—of various colors, shapes, and sizes, from the three-cornered batteau, scarcely capable of bearing one man, to the small steamship that carries 320 head of cattle to Cuba. If the day is favorable, we will likely see a number of white sails up the river and down the bay," wrote one visitor.[20] Looking back, Tampa newspaperman D. B. McKay recalled that his "Grandpa [James] McKay's schooners brought a wide variety of goods to Tampa in the early eighties. I remember many a train of wagons drawn by oxen hauling goods brought in by these schooners departing for the wilderness of south and central Florida."[21] Nevertheless, Tampa Bay remained a backwater compared to Pensacola, Cedar Key, and other Gulf ports.

Shallow water was to blame. There was "plenty of bay to look at," one visitor noted, but "not to sail in."[22] Another was blunter: "Tampa . . . has not a first class harbor."[23] The U.S. Army Corps of Engineers, the federal body charged with developing and maintaining the nation's ports and navigable waterways, sent an engineer to survey the bay in 1871. He proposed cutting a channel through Hillsborough Bay, but his superiors disagreed and refused to pursue the project. Eight years later, engineers returned and found that silting, perhaps from increased runoff spurred by land use changes in the watershed like logging and cattle raising, was making Hillsborough Bay shallower north of Ballast Point. At only five to eight feet deep, it was dangerous even for shallow-draft boats. In 1880, Congress allocated the funds to dredge a nine-foot-deep, 150-foot-wide channel to the mouth of the Hillsborough River. It was the first major alteration of the bay bottom. It required constant maintenance.[24]

This small change paled next to the dramatic destruction of the bay's wildlife facilitated by the growth in trade. In the late nineteenth century, feather-covered hats were the height of women's fashion. The plumes of wading birds were especially popular. Florida, home to millions of the most desired birds, suffered an invasion of commercial hunters who, armed with small caliber rifles, would approach a rookery by water and shoot them out as the birds nested.[25] Tampa Bay was an easy and popular hunting ground. The variety and numbers of wading birds astonished ornithologist W.E.D. Scott when he first visited the Bay Area in 1880. When he retraced his footsteps in 1887, however, he found that hunters had decimated nearly all the bay's rookeries. A visit to Double Branch and Rocky Creeks at the northern end of Old Tampa Bay, where he had "once seen birds so abundant," left him in despair. "Formerly I had seen birds breeding here in great numbers, and Reddish Egrets had been the most conspicuous feature. . . . But now how different! Not a single pair of birds of any kind did I find nesting, and only at rare intervals were any kinds of Herons to be observed." Searching Terra Ceia Bay on a hint from local fishermen that a large rookery could still be found there, he saw nothing but a "half a dozen Herons and about fifty White Ibises, all of them flying so far back into the interior that I lost sight of them. If there was any large rookery on this shore I was unable to find it."[26]

According to Scott, the ground zero of plume hunting was the Maximo Rookery, a massive mangrove island in Boca Ciega Bay, where Alfred Lechevalier led a team of plume hunters from his camp on what is now Frenchman's Creek. Scott had visited the rookery before Lechevalier's arrival and found that it "teemed with bird life." "Every tree and bush . . . contained at least one nest, and many contained from two to six or eight nests. . . . A perfect cloud of birds were always to be seen hovering over the island in the spring and early summer months, and conspicuous among them were Brown Pelicans, Man-o'-war Birds, Reddish Egrets, Florida [Double-crested] Cormorants, Louisiana [Tricolored] Herons, American [Great] Egrets, Snowy Herons, Little Blue Herons, Great Blue Herons, and both kinds of Night Herons." According to Scott's local informants, Lechevalier and his hunters took only "five breeding seasons to break up . . . this once incomparable breeding resort," before moving on to other parts of the state.[27]

The casualness with which people killed birds astonished observers. A "ruthless and wholesale massacre of birds is being carried on down the bay," one correspondent wrote in 1900, explaining that "several camps of men" had "set fire to one of the small islands which was the nesting place of thousands of birds."[28] In 1894, a child from Palma Sola Bay wrote of trips to hunt pelican eggs on Passage Key, at the mouth of the bay: "It is such fun to hunt for them, as the pelicans build their nests right on the ground."[29] Fishermen, seeing pelicans as competition, killed them with little thought.

The assault on nature was not limited to birds. Florida's natural environment endured a multipronged onslaught beginning in the late nineteenth and early twentieth centuries. Hunters shot nearly anything that could provide a skin or fur. Naval stores producers, especially turpentine makers, eyed Florida's vast pine forests for their untapped potential. Logging companies did the same, taking pines, hardwoods, and cypress. Sawmills churned out millions of feet of boards and shingles. Drainage projects turned wetlands into dry ground, while boosters pushed to make the dry ground into farms. Miners stripped back the land for the phosphate that lay beneath it. Railroads extended the reach of the market's hand. Crews built new lines across Florida in the last decades of the nineteenth century. On the Atlantic Coast, Henry Flagler's Florida East Coast Railroad, with its tony resorts and glittering new cities, grabbed most of the glory. But the railroads that crossed the state's interior did the dirty work of hauling commodities to faraway places where lumber became houses, alligators became leather, and phosphate became fertilizers. When the South Florida Railroad (later the Plant System) reached Tampa in 1884 and the Orange Belt stretched down the Pinellas Peninsula to what became St. Petersburg in 1888, they replaced the rutted wagon roads and cow paths that once trailed into the hinterland and connected Tampa Bay to all the action.

It is hard to overstate the importance of the arrival of the railroads to the Bay Area. Tampa went from sleepy backwater town to bustling small city, especially after the establishment of cigar factories in Ybor City in 1886 and West Tampa in 1895. St. Petersburg, meanwhile, owed its very existence to the extension of the Orange Belt Railroad. Settlers and tourists rode the railroad to the mineral spring at Safety Harbor on the northwest shore of Old Tampa Bay. Followers of the ideas of Utopian thinker John Ruskin established a town named in his honor along the Little Manatee River just a few miles from the eastern banks of Hillsborough Bay. Between 1880 and 1910, the population around the bay more than doubled every decade.[30]

The railroad's arrival created new markets for the bay's commercial fisheries. Americans' taste for fish had changed by the late nineteenth century: salted fish was out, fresh fish, delivered via iced railcars, was in. While a ramshackle operation, the Orange Belt Railroad put St. Petersburg on the map in the fish trade. North Carolinian Henry W. Hibbs opened a fish house on the Orange Belt's pier when it opened in 1889 and quickly attracted fishermen working between Old Tampa Bay and Boca Ciega Bay. Over the next decade, he and others who followed his lead shipped millions of pounds of snapper and mackerel a year north from St. Petersburg.[31]

Mullet, bountiful and cheap—eight cents a pound in Atlanta—and a favorite among working people across the South, was the bay's most popular catch.[32] In the 1880s, mullet fishermen from eastern North Carolina established the town of Cortez on Hunter's Point, just south of the mouth of Tampa Bay. They fished all around the lower bay using techniques for catching mullet that had been perfected on North Carolina's Pamlico Sound. They hauled their catches to Tampa for shipment north. "The amount of fish carried away . . . each week," one observer wrote from

Cortez in 1897, "have at no time fallen below thirty-five thousand, and had frequently gone to sixty-thousand and over. A low estimate would be two hundred thousand fish each month, which puts in the pockets of the fishermen about twenty-five hundred dollars."[33]

Tampa-based fishermen competed for the catch. Where the Cortez fishermen worked mostly with small boats crewed by family members, many Tampa fishermen worked for companies with small fleets of boats. John Savarese owned dozens of fishing boats on which he employed hundreds of fishermen who fanned out around Tampa Bay and all along the southwest coast in search of mullet, pompano, and snapper. His "principal camp is at Hunter's Point, where some sixty men are constantly employed," a reporter noted in 1897. "Twenty men are also employed at each of the following camps: Little Manatee river and Old Tampa bay, while ten to fifteen expert hands keep things moving at Picnic Island, Cedar Point, Sarasota, Lemon bay, Stump Pass, and Cockroach bay." His several steamers carried the catch back to Tampa for shipping.[34]

Tensions between Cortez and Tampa fishermen were common. Cortez fishermen complained that large Tampa firms used unscrupulous and illegal methods to catch the mullet as they exited the bay. But battles were often about more than methods or claims to space. Racism against the Black fishermen employed by Savarese and others shaped the competition. In 1895, gunmen fired thirty shots at a Tampa-based boat as it passed Hunter's Point on its way back up the bay. "The guilty parties are supposed to be certain North Carolina fishermen who [had] taken umbrage at the employment of colored fishermen by Mr. McIlvane," the *Tampa Tribune* reported.[35] Two years later, a white Cortez fishermen complained of the "Tampa fishermen, with their thousands of yards of stop net [and] their negro crews," who "invade our waters" to take the fish.[36]

By the 1890s, overfishing had reduced the size of mullet catches in and around Tampa Bay. In 1897, the state passed its first conservation law. The law established a protected season for mullet and banned practices like stop netting. Enforcement was lax at best, owing to the minimal expenditures the state provided for wardens to enforce the law. "No attention whatever is being paid to the fish laws of the state," the *Tampa Herald* reported in January 1905. "Mullet are being caught and sold in the streets of Tampa, and the famous 'stop net,' which the laws have long since put out of business, is reported as being used along the coast."[37] Mullet fishing had its ups and downs but remained an important industry through the end of the twentieth century, when a 1994 amendment to the Florida constitution banned the use of gill nets—the industry's most important tool.

Oystermen joined the fishermen in harvesting the bay's resources for far-off markets. In the 1890s, oystermen shipped hundreds of thousands of pounds annually.[38] One journalist noted that R. W. Warner & Co., one of the leading oystering companies, shipped about forty barrels of oysters in the shell and another sixty to eighty gallons of shelled oysters a day, in addition to those they sold right off the boats at its Port Tampa restaurant. "Employment is given to about forty men, all the time, who use steam yachts, sailing yachts and row boats to carry on the business." They collected oysters from bars on every part of the bay, but those from Rocky Point on Old Tampa Bay were considered the largest and best. Oystermen "planted and transplanted and worked upon" the beds for years "to bring to the highest state of the most delicious maturity."[39] As with fish, overharvesting took a toll on the business, but pollution proved more troublesome in time. By the 1920s, effluent from Tampa's sewers led health inspectors to close the beds of McKay and Hillsborough Bays to oystering; Old Tampa Bay retained a viable, if much reduced, oyster industry for several decades more.

While fishermen and oystermen shipped their products inland by rail, others hoped to use the railroads to drive freight to the sea. Tampa Bay's proximity to the state's burgeoning groves, logging camps, and phosphate mines seemed to make it a perfect place to ship these bulky cargos. But one hang-up remained: the shallowness of Hillsborough Bay. As steamships grew in the late nineteenth century, they needed deeper harbors. Hillsborough Bay, with its nine-foot channel, could not do the job. Henry Plant was the first to reckon with this fact. He initially built his rail line to stop at the banks of the Hillsborough River, but his steamships could not reach the river. Frustrated, Plant extended his railway line nine miles to Passage Point on Old Tampa Bay in 1888, where the water reached a natural depth of just over fifteen feet. He renamed it Port Tampa. The Plant System's massive dock featured a mile-and-a-half-long railroad pier to allow the trains and ships to meet. That same year, Congress allocated funds to dredge a twenty-foot-deep channel to the pier. Dredgers completed their work five years later. The dredged material from the bay bottom was used to form Picnic Island. Meanwhile, the channel in Hillsborough Bay became two feet shallower as maintenance dredging came to a halt.[40]

The pier at Port Tampa bustled. Freight and people poured in and out as ships and trains came and went. "It was an odd sight watching the men at work, mostly black men, under the electric lights, with their songs and shouts, getting into the hold a quantity of timbers," an observer reported. "The pier was covered with trunks and cases, passengers of all sorts and nationalities crowded on board, and enough bags of mail were stowed away somewhere to keep all the people of Key West and Havana busy for a week reading their letters."[41]

Port Tampa, along with the smaller port at St. Petersburg that opened when the Orange Belt Railroad completed its railroad wharf in 1889, established Tampa Bay as a zone of ecological transfer. Longshoremen loaded lumber by the board, phosphate by the shovelful, and oranges by the crate onto ships that sailed to ports all over the world. At the same time, ships jammed with leaves of Cuban tobacco and Honduran cedar fed the growing cigar industry. In time, schooners loaded with coconuts and bananas unloaded their tropical treasures on the bay's wharfs. Plant's steamers, the *Olivette* and the *Mascotte*, which plied the Gulf of Mexico between Port Tampa, Key West, and Havana, also carried generations of Cuban and Spanish migrants and immigrants to work in Tampa's cigar factories. In 1892, a Congressional commission found that upward of 100,000 people traveled between Cuba and Florida every year, many of them to Tampa.[42]

Increased shipping and, especially, the close ties to Cuba were not universally cheered. For white Tampa Bay residents, the links suggested not only commerce but danger, particularly from yellow fever. Caused by a virus spread by the *Aedes aegypti* mosquito, yellow fever struck Tampa periodically before the 1880s. Political leaders suspected the increased trade was the cause when a particularly virulent epidemic pummeled Tampa in 1887. They were not entirely wrong: the disease traveled as mosquitos carried the virus from one infected person to another. Infected persons moving about on ships or infected mosquitos trapped in the holds of ships could move disease from place to place. Before the discovery of this pathway, though, many Americans assumed it was a product of unclean living conditions, and they blamed the Cuban travelers. Local leaders suspected all boats coming from Cuba could be carrying the disease. They also blamed Cuban sailors who used small boats to smuggle *aguardiente* to Tampa drinkers. At the request of local officials, federal authorities established a quarantine of the port to prevent all but a few commodities, particularly tobacco leaves, from entering. They also banned Cuban smacks from entering the bay, although many continued to carry out business off-shore. Ultimately, the federal government established a quarantine station at Egmont Key, where boats and passengers suspected of carrying yellow fever were disinfected before entering the bay. This surveillance regime remained in place until well into the twentieth century.[43]

The Cuban connection turned the federal government's attention to Tampa Bay again in 1898 with the declaration of war against Spain. Enticed by Henry Plant's promises that his port would be the ideal place from which to launch troops to Cuba and Puerto Rico, the War Department chose Tampa as the point of departure. Almost immediately, it was chaos. By the 1890s, a second railroad company, the Florida Central & Peninsular Railway (later part of the Seaboard Air Line), had begun to serve the city of Tampa, but only Plant's railroad served Port Tampa, and he refused to allow any other railroad access to his track. He even continued civilian passenger service to Port Tampa in the middle of the mobilization. The port simply lacked the facilities to handle the volume of traffic. Bottle-necks snagged the loading of supplies. Soldiers often waited days on board before the ships could even head to sea. The army created many of its own problems, but the whole affair made Port Tampa's deficiencies clear.[44]

Tampa's leaders used the debacle to push for port facilities along the city's waterfront. Frustrated by the costs of using Plant's pier on Old Tampa Bay while they stared out on a wide, shallow, and largely empty Hillsborough Bay, city leaders begged the Army Corps of Engineers to dredge a new channel to downtown. The ascendance of one of their own, Congressman Stephen M. Sparkman, to the House Committee on Rivers and Harbors, gave them the power to turn their pleas into action. In 1899, Congress allocated the funds to deepen the Hillsborough Bay channel to twelve feet, while deepening Port Tampa's channel to twenty-seven feet. Six years later, Sparkman delivered again, landing a new allocation for the completion of a twenty-foot-deep channel straight up the center of Hillsborough Bay. By 1907, the work was largely done.[45]

Hoping to take advantage of the new channel, railroads and real estate developers dug out their own channels to remake Tampa's waterfront. The Seaboard Air Line dredged the Seddon Channel on the west side of Big Grassy Island. Renamed Seddon Island, it was the location of the Sea-board's phosphate port until the 1970s. Developers renamed it Harbour Island in the 1980s. The Tampa Northern Railroad, which extended its

Figure 3. *Top*, The Estuary, 1911. *Bottom*, Ybor Channel, 1949. Making Hillsborough Bay into a viable port required the dredging of ship channels and docking spaces. The most radical change was in the area known as the Estuary, which was renamed the Ybor Channel after it was dredged. The top photograph shows the Estuary viewed from the north shortly before dredging of the Ybor Channel began. The building on the right side of the image is the Tampa Gas Works, which was located on what is now the northwest corner of the intersection of Channelside Drive and Adamo Drive. The bottom image shows the same area viewed from the south on Hooker's Point. The Army Corps of Engineers used the spoil from the channel dredging to build up the dock space on the west side of the channel. The cylindrical building near the center of the image is the Tampa Gas Works site. Credit: Burgert Brothers Photographic Collection, courtesy of Tampa-Hillsborough County Public Library System.

tracks to Tampa in 1906, dredged the island's east side, cutting what is now the Sparkman Channel. Hendry & Knight, the real estate firm that controlled the city's waterfront, dredged north of Seddon Island, creating the Garrison Channel, on which they built a series of warehouses. Recognizing that the dredging of Hillsborough Bay was likely to benefit only those few private interests, the Army Corps of Engineers suggested the wetland east of Tampa known as the Estuary be dredged to a width of three hundred feet to allow for docks the city would control. In 1910, Sparkman, then chairman of the Committee on River and Harbors, delivered the money, but it took more than a decade to complete the Ybor Channel and the municipal dock. The Army Corps of Engineers eventually took over management of the complex of channels, and by 1928, had dug them to a depth of twenty-seven feet.[46]

Calls for deepening, widening, and extending the bay's shipping channels renewed almost as soon as the dredges finished each project. Ships were getting larger, and the weight of the heavier loads, particularly of phosphate, made them sit lower in the water. By the late 1920s, port managers reported dozens of groundings, bottom strikes, and near collisions. In 1933, prodded by the pleas of shippers and the Chamber of Commerce, Congress approved deepening the ship channel to thirty feet and widening it to three hundred feet.[47]

Dredging forever changed the shape of the bay's bottom. Naturally, the bay's six-foot contour extended from most shorelines for a couple miles before sloping to deeper water that occasionally might drop to twenty-five feet but averaged closer to ten feet. As dredgers sucked up the shallower sandy bottoms and blasted through the limestone base, they destroyed sea grass habitats and left deeper, darker water that altered aquatic habitats at the center of the bay. Dumping the dredged sand on shallow seagrass beds and salt marshes introduced equally damaging changes. The dredgers of the Sparkman Channel piled sediment on Hooker's Point to add space for docks and in the process destroyed its natural shoreline and narrowed the flow of water into and out of McKay Bay. Digging the Estuary to make the Ybor Channel was probably the starkest picture of destruction. A "big grassy sawgrass flat" before dredging, it was, a dozen years later, the site of "streets, tracks, [and] buildings" destined to become "Tampa's industrial and shipping center."[48]

City leaders at the time cared little about the ecological costs of building the port. Dredging Hillsborough Bay brought the development advo-

Figure 4. Ships docked in Hendry and Knight Channel, 1926. By the 1920s, Tampa was one of the leading ports on the Gulf Coast, with ships coming from and going to points all over the world. After the dredging of Hillsborough Bay, the docks along Tampa's waterfront supplanted those at Port Tampa as the center of port activity. The Hendry and Knight Channel, today known as the Garrison Channel, offered dock space on both sides, as this image, looking east from the Hillsborough River, shows. Credit: Burgert Brothers Photographic Collection, courtesy of Tampa-Hillsborough County Public Library System.

cates had promised and that was enough for them. Port traffic increased. Shipping companies built new warehouses along the Garrison Channel, oil and concrete facilities on Hooker's Point, and a phosphate terminal on Seddon Island. The city, meanwhile, erected a massive municipal warehouse along the slips in the Estuary. By 1927, sixteen wharves covered more than two-and-a-half miles of waterfront.[49] By the early 1930s, Tampa was the nation's seventeenth busiest port, served by fifty-six steamship lines that carried its cargoes all over the world.[50]

A journalist described the scene along Tampa's waterfront in 1935. As the Standard Fruit Company's refitted naval destroyer arrived with bananas from Cuba, a schooner, "its hold and decks burdened with lumber," sat low in the water preparing to retrace its path to "that long, winding line of islands in the West Indies and far up a tropical river." On another ship, "passengers [stood] on the bridge with the master as a big ship moved slowly down the channel. At the dock, one sailor "set his foot on solid ground for the first time in a month," while another sat "in a scaffold and daubed red lead" into the hull of his rusting ship. Other sailors "danced with the girls in a 'waterfront joint' to the blaring music of a slot phonograph." Meanwhile, African American longshoremen played bottle cap checkers as they waited for a ship to unload. Their ships tied up "stern to prow," captains from as far away as Yokohama and Hamburg "renewed an old acquaintance made in Cape Town."[51] It was quite a change for a port that less than a half-century earlier had no vessels running directly to foreign ports.[52]

Imports, particularly bananas and coconuts from Central America and the Caribbean, tobacco from Cuba, and cedar logs from Honduras (the latter two critical for the cigar industry), linked Tampa Bay to the Americas, but exports connected it to the world. Refrigerated ships took oranges and grapefruit as far away as London. Lumber ships carried millions of board feet of lumber taken from Central Florida's pine forests to "the River Plate [Buenos Aires], the north coast of South America, Panama, Cuba, the West Indies, Spain, Italy, England, Holland, Germany, France and West Africa."[53]

All other exports paled next to phosphate, however. By the 1920s, phosphate accounted for 90 percent of the bay's export traffic.[54] Deposited during the peninsula's millions of years under ocean water during the Miocene, phosphate is an essential component for making chemical fertilizers and thus the key to increased agricultural production worldwide in the twentieth century. A mining boom followed its discovery in Central Florida in the 1880s. Production centered east of Tampa Bay in a region known as Bone Valley that stretched from Hillsborough County east to Polk County and south to Manatee and Hardee Counties. By the 1920s, miners pulled millions of tons of phosphate from the ground every year. Freighters carried nearly all of it through Tampa Bay to as far away as Japan and Australia and to points all over the globe. In 1929, the U.S. Phosphoric Corporation received permission to dredge a three-mile-long, twenty-four-foot-deep channel from the main channel to the mouth of the Alafia River, where it built a phosphate processing plant.[55] The mound of acidic and radioactive waste products created by the plant, commonly known as a gypsum stack, has become a landmark on the bay's eastern shore. As Tampa's modern port facilities took shape, the demands of the phosphate industry did much of the shaping.

Figure 5. SS *Edith* loading phosphate at Seddon Island elevators, 1957. Phosphate has been Tampa Bay's most important export since the early twentieth century, and the arrangement of the port reflected the needs of phosphate shipping. While railroads continued to carry phosphate to Port Tampa well into the twentieth century, the busiest elevators for loading ships were on Seddon Island. After World War II, increased production meant more and larger ships. In March 1957, the Bull Steamship Company's SS *Edith* took on a load of 10,500 gross tons in its five holds. The *Tampa Tribune* reported that it was the largest load ever sailed from Tampa up to that point. The crew sailed the cargo to New Orleans, where it was used to make fertilizer. In this image, airborne phosphate dust escapes from the ship as the hold is loaded. Airborne pollution from the phosphate docks contributed to high levels of phosphorus in the water around the port. Residents on Davis Islands regularly complained about the accumulation of phosphate dust on their yards. Credit: Burgert Brothers Photographic Collection, courtesy of Tampa-Hillsborough County Public Library System.

World War II spurred further transformation of Tampa's harbor facilities. At Hooker's Point, which grew longer and wider every time dredgers needed a place to dump bay bottom, two shipyards sprang up. At the Tampa Shipbuilding (TASCO) yard, 16,000 workers launched 494 ships of various sorts during the war. Next door, McCloskey Shipbuilding churned out concrete ships in three massive concrete-lined basins dug into the shoreline. Meanwhile, workers piled into six hundred new houses in the Maritime Homes development built by the U.S. Maritime Commission.[56]

McCloskey did not survive the end of the war and TASCO shrank in size as government contracts evaporated, but Hooker's Point grew as an industrial zone and became the center of Tampa Bay's port facilities in the postwar years. By 1950, more than a dozen manufacturers, importers, and a citrus cannery operated on the growing spit of land. A large grain elevator, an oil refinery, and an electric generating plant followed.[57] McCloskey's old basins became the home of Tampa's shrimping industry. The Singleton Shrimp Company landed the first trawler, loaded with 150 boxes of shrimp from the Campeche Banks off the Yucatan, in 1951.[58] Three additional shrimp processors followed, employing more than a thousand workers, mostly women, who peeled and deveined, butterflied, and breaded shrimp for a nation newly enamored with frozen seafood. At the industry's height, more than two hundred trawlers supplied the processors with shrimp caught from across the Gulf of Mexico and as far as South America.[59] Adjacent to the shrimp docks was the nation's largest turtle processing facility where, until the 1960s, workers processed tens of thousands of green turtles caught in the Caribbean into steaks, soup, and myriad industrial products. While Port Tampa remained a busy petroleum port, shipping companies abandoned the warehouses that once snaked along the east side of the Hillsborough River for the east side of downtown.[60] Phosphate and petroleum, citrus and cement moved through Hooker's Point, which even boasted the state's only deep-water facility for shipping cattle. William Hooker might have recognized at least that much.[61]

The filling in of the east side of McKay Bay meanwhile brought its shoreline closer to Hooker's Point. As early as 1926, planners had envisioned filling in northern arm of McKay Bay to make it a port.[62] Those ideas came to fruition between the 1950s and 1970s as new port projects reduced McKay Bay's size by half. In 1954, the Port Sutton Company announced plans to develop the spit of land created when the U.S. Phosphoric Company dredged its channel. By the early 1960s, they had built up 150 acres by dumping dredged materials on the shoreline.[63] Ships loaded with coal for Tampa Electric's generating plant docked there alongside ships ready to carry phosphate out. In 1968, the Seaboard Coast Line filled more of the shoreline north of Port Sutton and built new phosphate elevators to replace those it operated at Port Tampa and at Seddon Island. When the company finally abandoned the old elevators in the mid-1970s, Tampans, frustrated by the seemingly interminable phosphate trains passing through downtown, rejoiced. In the early 1970s, the Tampa Port Authority, a public agency created in 1945 to oversee the port's operation and expansion, filled in the remaining mangrove shoreline and adjacent shallows to make room for the East Bay Terminals, a modern dockage facility built to replace the aging warehouses along the Garrison Channel.

Port growth expanded in other parts of the bay as well. Port Tampa handled nearly half of the bay's petroleum traffic, including fuel deliveries for MacDill Air Force Base. Florida Power maintained a terminal for fuel oil deliveries at Weedon Island. Bayboro Harbor in St. Petersburg, while comparatively small, had a deep-water channel capable of handling large ships. In 1970, Manatee County added to all this when it opened Port Manatee in hope of luring phosphate shippers away from the ports in Tampa.

A cycle set in during the postwar years: deeper and wider channels provided more land for port development while larger ports depended on attracting larger ships, which, in turn, necessitated wider and deeper channels. In 1955, the Army Corps of Engineers began a four-year project of deepening the main ship channel, the Port Tampa Channel, and the channels and turning basins of upper Hillsborough Bay to thirty-four feet. By 1966, Sam Gibbons, Tampa's longtime representative to Congress, requested that body to allocate the funds to deepen those channels to forty feet.[64] Three years later, supporters of the port began advocating for a main channel depth of forty-two feet.[65] In 1970, Congress agreed to a forty-three-foot-deep channel. "It will be the best investment this area ever made," claimed the chairman of the port authority.[66]

Few in power, from local politicians to business leaders to government engineers, doubted the logic of growth in the decades after World War II. "The need for deeper channels in Tampa Bay and Harbor stems directly from our American way of life," an officer at the Army Corps of Engineers wrote in 1946, capturing the optimistic spirit of the postwar years. "Our national economy continues to demand bigger, better and faster units of transport."[67] To those charged with overseeing the economic well-being of the region, bigger, better, faster made sense. The port created jobs—one in ten in Hillsborough County, the Tampa Maritime Association claimed in 1970—and provided the rising tide of economic growth that, at least proverbially, lifted all boats.[68] The ecological costs, however, never showed up on those ledgers.

4

PLAYGROUND

Henry B. Plant built the Tampa Bay Hotel to be the star attraction at the end of his railroad. Completed in 1891 at a cost of $2.5 million and appointed with a half-million-dollars' worth of artwork and furniture, the hotel epitomized Gilded Age luxury. Rooms provided guests comfortable accommodations, beautiful decorations, and had the newest technologies, including electricity and telephones. Overstuffed couches invited visitors to converse in view of fine statuary in the great hall before partaking of sumptuous meals in the dining hall. An orchestra played in the music hall; leading actors and singers entertained large crowds in the Tampa Bay Casino. Above it all stood the crescent moons on the minarets that epitomized the hotel's mixing of Moorish and Byzantine styles.

If the hotel was the star, the environment was a very powerful supporting cast. The beguiling winter sunshine of Florida's Gulf Coast lit the production. Languid, Spanish-moss-strewn live oaks conjured romantic images of southern latitudes for Plant's prospective visitors. "Each room is an outside room," Plant promised visitors. "All of them admit the sunlight and the refreshing and delicious salt breezes."[1] To the natural treasures, Plant added gardens in which to stroll and orchards from which to pluck oranges. A golf course merged sport and nature. The hotel employed guides for hunting trips into surrounding woods and fishing expeditions down into Tampa Bay in search of the pompano, tarpon, Spanish mackerel, and "many other celebrated fish."[2] An ornate boathouse on the west bank of the Hillsborough River provided a variety of small boats for guests to sail along the shore of Hillsborough Bay.

In building his hotel, Plant hoped to make Tampa Bay a playground for the wealthy akin to that being built by his friend and competitor Henry Flagler along the Atlantic Coast. But Tampa Bay was no Palm Beach. The famous and wealthy visited, especially after the attention the hotel received as the headquarters of U.S. Army forces deploying for the Spanish-American War, but they bought up beachfront property and built stately mansions where Flagler's, not Plant's, railroad deposited them.

In the long run, this turned out for the best. Unlike the exclusive beach communities of the east coast, Tampa Bay offered plenty of space for people from all walks of life to swim, fish, sail, and enjoy the shore. Part of this was a function of the sheer size of the bay, of course, particularly when the region's population was still relatively small. But in Hillsborough and Pinellas Counties, a democratic impulse to maintain public access shaped how people encountered the bay's waters. Public beaches, fishing piers, parks, and boat ramps made recreational access easy. In some places, all you had to do was pull off the road and you could swim, sun, and fish to your heart's content. It was not true in all circumstances. Laws barred Black people from many public places, and class played a role in determining where and in what ways people played on the bay. But average folks made it clear that Tampa Bay was theirs. The guy putting in his boat on the Courtney Campbell Causeway, the woman crabbing with a chicken leg quarter at the Little Manatee River, the midnight fisherfolk on the Gandy Bridge, the kiteboarder setting off at Skyway Beach, the college students soaking up rays at Spa Beach, the kids splashing in the water with their dog at Picnic Island, and the family throwing a birthday party at DeSoto Park: the bay belonged to all of them.

Private development nevertheless ate up space around the bay, slowly at first, quickening during the land boom of the 1920s, and accelerating rapidly in the years following World War II as the American Dream came to include a place in the sun and a boat on the water. Hotels, condominiums, and houses rose around the bay as land developers built the Florida Dream on man-made shorelines. Causeways and bridges tied it all together but at great environmental costs. The parceling of the bay's shoreline restricted access while dredge-and-fill practices turned submerged public lands into dry private property. As happens in any contest for a resource, private and public interests collided often.

Neither public use nor private development held a monopoly on the environmental consequences of development for Tampa Bay. But neither were their consequences equal. Providing usable access for the public required infrastructure. Seeing the water is one thing; accessing it comfortably and playing by and in it safely was more complicated. Piers had to be extended, marinas had to be dug, and beaches had to be built. All of those projects took a toll on the shoreline. But the massive dredging projects that built places like Davis Islands, Beach Park, Venetian Isles, Apollo Beach, and the dozens of other communities around the bay changed things immeasurably. Tampa Bay was the selling point that drew people to the area. But it was not the pristine, secluded bay visitors and new arrivals imagined. The development of the port had guaranteed that, but so, too, did the very act of making it a destination.

Fishers, not sunbathers, were among the first to seek out Tampa Bay for recreation. In the late nineteenth century, sportfishing became all the rage among the nation's elites and growing middle class. Florida's Gulf Coast was on top of the lists of places to cast a line. "We know of no fishing in America equal to the Gulf coast from the bight in the shore line above Cedar Keys down to Key West," a correspondent wrote in 1888.[3] From dinner plate delights to hard fighters, the region's piscatorial bounty lured anglers. The most popular target was tarpon. The "Silver King" was no good for eating but gave a fight like few others. The epicenter of the tarpon hunt lay to the south around Charlotte Harbor, but Tampa Bay was a hot spot, too.[4] The mouth of the bay, from Pass-a-Grille across to Longboat Key, was the best place to find them, but tarpon could be found anywhere smaller fish congregated. Reports of tarpon being caught off the docks on Hillsborough Bay or even in the lower Hillsborough River were not uncommon.

Proper sportfishing required gear and a guide. Most tourists went the cheaper route and fished from the shore. The Orange Belt Railroad pier, which stretched 3,000 feet into the bay from its base in downtown St. Petersburg, hosted more fishing parties than it ever did railcars. "The first thing nine out of ten tourists do when arriving at St. Petersburg is unpack or buy . . . fishing tackle and head for the bay," joked William Straub in 1898.[5] When the pier changed hands in the 1890s, the fishermen remained. When it was torn down and replaced by the Electric Pier in 1905, the fishermen moved to the new one. A similar scene could be found on Plant's pier at Port Tampa. The variety—sheepshead, trout, jacks, mackerel, kingfish, grouper, redfish, snappers, and more—and their abundance made fishing easy. "No special directions are needed when the fish are running

Figure 6. Mrs. Joseph White and son with tarpon, 1925. Tarpon fishing has been a popular pursuit in and around Tampa Bay since the late nineteenth century when adventurous game fishers came in pursuit of the Silver King. While often depicted as a pursuit for men, women and children also engaged in the sport. Though tarpon may have gotten top billing, recreational fishers pursued a wide variety of game fish and food fishes around the bay. Credit: Burgert Brothers Photographic Collection, courtesy of Tampa-Hillsborough County Public Library System.

in schools," one sportsman noted, "except to keep the bait or fly in constant motion on the surface—the fish will do the rest."[6]

Good fishing could be had almost everywhere around the bay, which was a boon to the region's locals as well as tourists. The shoreline was a diverse space. White folks, Black folks, Cubans, Spaniards, and Italians, people of all economic classes could fish the bay just about any time they wanted. Many caught their dinners along the shoreline. "Lots of people would come out of Ybor City with a cane pole and they would dip shrimp [use a dip net to snag bait shrimp]" to catch sheepshead and mangrove snappers, one resident recalled.[7] Others made a living selling their catches to fish dealers and restaurants. They also picked oysters, hunted stone crabs, grabbed scallops in shallow seagrass beds, and sometimes dug coquinas out of the sand for soup. Blue crabs were a common sight on the dinner tables of the working classes. Cigar workers fed their families with them during strikes. Tampa's famous deviled crab—made from a mixture of crab meat, stale Cuban bread, and spices—is a result of this tradition of eating from the bay. "We can take a hook and line and catch a mess of fish," one writer boasted, "or gather oysters by the train load, and stone crabs, and blue crabs, and laugh at the price of pork chops."[8] People around Tampa Bay would have heartily agreed. Fishing and crabbing were social occasions, too. People of all backgrounds, from "popular Hyde Park girls" to Cuban social clubs to Black church groups held crabbing parties.[9]

Swimming, pleasure boating, and collecting shells also became popular pastimes as ideas about people and the sea evolved in this era. Romantic attachments to the sea's beauty replaced fears of it as a menacing hazard to be crossed quickly. Shorelines became playgrounds on which to frolic in the sun. Sea life lost its terror, appreciated instead for its strange beauty. Doctors prescribed exposure to salt water and ocean air as a cure for all sorts of maladies, especially those brought on by life in the nation's mushrooming urban places. If Florida, surrounded by salt water on three sides, was the perfect place to expose oneself to the miracles of the sea, how much more were the peninsulas within the peninsula formed by Tampa Bay?

Entertainment facilities popped up around the bay as the population grew. Picnic Island opened in 1890 just south of Henry Plant's Port Tampa Pier. Plant's engineers created the beachfront when they dumped the spoil from dredging the channel. Plant then built an amusement park to attract Tampans to ride his railroad. It featured an outdoor bowling alley, a pavilion with a dance floor, and a restaurant that served "fish and clam chowders" to complete the seaside vibe. The "chief attraction" was the two-hundred-foot swimming area, dug out to a maximum depth of six feet and fenced to keep children from venturing too deep. Bath houses offered swimsuits for men, women, and children and freshwater showers for rinsing off in privacy. Rowboats, sailboats, and other launches could be rented as well.[10] In 1894, Tampa Electric's streetcar company extended a line to Ballast Point where Emilia Chapin, the line's controlling owner, built a two-story pavilion that looked out over Hillsborough Bay. In addition to a restaurant that served "shore dinners" and a theater, the pavilion also had a "bathhouse where bathing suits could be rented" and a "daring two-story toboggan slide into the water." There was also a Ferris wheel and a playground.[11] That same year, the Tampa and Palmetto Beach Railroad Company, a competitor of Tampa Electric, donated land to the city at Palmetto Beach, at the northern end of McKay Bay, for the construction of DeSoto Park.[12] Less extensive than either Picnic Island or Ballast Point, it nevertheless had food stalls, a dance pavilion, a bandstand, a two-hundred-foot pier, and bath houses.[13]

Access to the bay reflected the official racism and the ethnic and class divisions of the region in this era. "For decent white folks only," read a sign spotted on a Tampa beach pavilion in 1898.[14] Picnic Island served the working classes of Ybor City and West Tampa mostly. Cuban, Spanish, and Italian mutual aid societies held picnics there. In 1899, a gathering of five hundred Black Tampans converged on the park for a large picnic on one of the days set aside for Black swimmers.[15] The park and pavilion at Ballast Point banned Black people from its beginning. The Tampa Electric Company, which controlled the park, allowed Cuban, Spanish, and Italian mutual aid societies from Ybor City and West Tampa to hold events like Labor Day and Sunday picnics there until the first decade of the twentieth century, when "objection came from the Americans, who declared they wanted free access to the Ballast Point grounds on Sundays."[16]

Development of St. Petersburg's waterfront proceeded differently. Some hoped that the city might follow Tampa's lead and allow commercial interests to develop it as a port. William L. Straub, editor and part owner of the *St. Petersburg Times*, meanwhile, successfully pushed for a public park along the bayfront. By 1916, Waterfront Park was "one of the largest public waterfronts in the nation."[17] The city dredged the shoreline, installed a seawall, and planted landscaping fit for a tourist dreamscape. In

1917, Straub and his allies in the Woman's Town Improvement Association made it clear that their goals for the park were largely aesthetic when they fought off a plan to lease the park to private interests for the development of amusements.[18] They did not wish to see Waterfront Park become a playground like Picnic Island or Ballast Point but, rather, remain a space from which serene views of the bay could be enjoyed.

Those looking for more active entertainment could go to the piers. The original railroad pier, in addition to excellent fishing, boasted a bathing pavilion and toboggan slide. The competing Brantley Pier, which operated from 1896 until it was replaced by the Electric Pier in 1906, likewise had a popular bathing pavilion. The Municipal Recreation Pier that replaced the Electric Pier in 1913 permitted visitors to drive their cars out over the water. It also featured a saltwater swimming pool, a spa, and Spa Beach.[19]

South of Waterfront Park, St. Petersburg's Black community made their own playground by the shore. Segregation laws barred Black folks from the Gulf beaches and Waterfront Park, but city officials allowed them to access the water at South Mole (now Demens Landing Park). The beach, a "rocky stretch of sand" left from dredging operations, had been used only as a place for the Atlantic Coast Line to park its idle cars.[20] South Mole lacked any facilities. Policemen often harassed bathers with spurious charges of indecent exposure, incidences of which, city officials admitted, were a result of a lack of changing facilities.[21] Just getting there required Black citizens to pass through neighborhoods where they endured the glaring eyes and jeers of white residents who disdained their presence.[22] City leaders heard loud complaints from white residents about the bathers but, mostly out of a desire to keep Black people out of more popular tourist areas, did nothing to bar them from the beach. Despite its limitations, South Mole became a special place for St. Petersburg's Black community. It was a place where children learned to swim, where ministers baptized believers. It remained one of the few Tampa Bay beaches open to Black bathers before desegregation in the 1960s.

Private development of the bay's shoreline expanded alongside public projects. Tampa and St. Petersburg went through a housing boom in the first decades of the twentieth century. As in many cities of the era, neighborhood development and streetcar expansion went hand in hand. New neighborhoods enticed the middle and upper classes to move away from working-class neighborhoods. Land developers looked to attract out-of-state buyers to purchase homes. Developers knew these buyers wanted land by the bay, so they subdivided waterfront land and started building improvements to make it more appealing.

Beginning in 1912, C. Perry Snell and his partners developed the area around Coffee Pot Bayou. They dredged the shoreline, installed seawalls, and dug out a channel from the bayou into the bay to "make one of the finest land locked harbors on the Florida coast."[23] Over on Hillsborough Bay, onto what was "once a paradise for fiddler crabs," the Swann and Holtsinger Company "pumped sand 24 hours a day" to build Suburb Beautiful, where, they assured, "breezes from the sea blow . . . , uncontaminated and unpolluted—fresh and always invigorating."[24] Developers of Sunset Park, on the Old Tampa Bay side of the Interbay Peninsula, dredged more than three-quarters of a mile into the coastline to "reclaim" thirty acres of land for homesites and "an ideal beach."[25] Advertisements promised there was "no finer place to bathe on Old Tampa Bay." Sloping gently from a seawall, the four-hundred-foot-wide beach had "hard, clean, white sand," a fine bath house, deep water for the daring, shallow water for children and the faint of heart, and "no under-tow."[26]

Real estate developers not only changed the bay's shoreline ecology, they challenged notions of development founded on the idea of the bay as a public resource. Snell and his partners envisioned that the Coffee Pot Bayou project would appeal to elites, who would "fill" the new marina with "private yachts and pleasure craft during the winter."[27] Swann and Holtsinger were more direct, promising that the neighborhood's "restrictions will prevent encroachment of undesirables."[28] It was, they assured white middle-class parents, "a children's paradise, nature's playground prepared and safe-guarded for their enjoyment, where all may enjoy the out-door life . . . in unconfined freedom with refined boys and girls."[29] In other words, they need not worry that Tampa's Black or Cuban children would be seen playing along the shore. Sunset Beach, meanwhile, was open exclusively to Sunset Park homeowners and their guests.[30]

Real estate developers and city planners alike thought their activities improved Tampa Bay. In 1921, nature pushed back on their hubris. It wasn't the most powerful hurricane ever to hit the Bay Area—the 1848 storm owns that title—but the 1921 storm was the worst in economic terms. (By some miracle, it remains so, but we'll deal more with that later.) Tidal surge, not the winds or the more than eight inches of rain, did the most damage that October 25th. The eye of the storm made landfall near Tarpon Springs. The counterclockwise winds of the storm's powerful east side tore a northerly path up Tampa Bay, pushing water up the bay until the tide reached 10.5 feet. The storm surge carried far inland all along the shore. In St. Petersburg, the storm destroyed the Municipal Pier. At Rocky Point, the storm swept away houses, and six feet of water pushed into the clubhouse of the golf course. A woman died after she and her husband clung to a cabbage palm through the night to avoid the rushing waters; the seventy-five-year-old man lost his grip on her when she lost consciousness, and her body was washed back into the bay. At Sunset Beach, houses floated away on the tide. The storm destroyed the beach at Picnic Island and the pavilion at Ballast Point. Water washed into the stately homes that lined Bayshore Boulevard. At Palmetto Beach, where there was no seawall, the tide pushed massive cedar logs—chained together and awaiting milling at the Tampa Box Company—inland and destroyed upward of fifty houses.[31]

But the hurricane of 1921 failed to chasten developers. The Florida land boom was in full swing, particularly in South Florida. Time spent in reflection was time lost to others peddling the Florida Dream elsewhere. Real estate salesmen and dredgers picked up where they left off. In St. Petersburg their work was especially obvious. According to historian Raymond O. Arsenault, "a five-mile stretch of Tampa Bay's natural shoreline was rendered unrecognizable" as "dredging and landfill projects turned shoals into channels and water into land, producing a series of artificial peninsulas that jutted into the bay."[32] On the Tampa side, developers carved canals into the Interbay Peninsula using dredge-and-fill techniques to make the exclusive suburb of Beach Park. "Residents are finding great sport in taking their speed craft or row boats and spending a few hours on the blue waters of the bay," touted the neighborhood's developers.[33]

The most audacious dredge-and-fill project of the land boom turned out to be David P. Davis's construction of the island that has borne his name since. Davis Islands, which boosters liked to tout as the as the "City that Emerged from the Florida Sea," started as Big and Little Grassy Keys, two small sedimentary islands at the mouth of the Hillsborough River.[34] Ignoring that they sat adjacent to the rapidly industrializing port—and right across the channel from the Atlantic Coast Line's phosphate elevator on Seddon Island—Davis was sure the location was a perfect place to construct a sort of Riviera on Hillsborough Bay. In 1924, after purchasing the islands and surrounding bay bottom, he hired workers to construct a concrete bulkhead around the islands and used dredgers to pump nine

Figure 7. Aerial view south end of Davis Islands, 1927. Building Tampa Bay's water-front communities depended on the sculpting of shorelines through the technique known as dredge and fill. The most audacious of such projects in this era was the construction of Davis Islands just south of downtown Tampa. Credit: Burgert Brothers Photographic Collection, courtesy of Tampa-Hillsborough County Public Library System.

million cubic yards of sediments behind it. In less than year, they had built up 875 acres to eight feet above sea level. Davis began selling lots immediately, and in due time real estate agents boasted of its tennis club, golf course, and, later, seaplane basin, as well as the beautiful homes with views of the "soft white-capped waves [of] Tampa Bay."[35]

Davis sold his development with all the ballyhoo worthy of the 1920s. Following the lead of his inspiration, Carl Fisher, developer of Miami Beach, Davis photographed bathing beauties, pulled publicity stunts, and staged events meant to attract the wealthy buyers he hoped would buy his property. Among these events were yacht races, a sport he believed would appeal to the elite buyers he sought. He proposed his first one in 1925, laying out a two-and-a-half-mile course west of Davis Islands and along Bayshore Boulevard. Making the course required dredging a two-hundred-foot-wide, six-foot-deep channel in that part of the bay.[36]

Davis was not alone in seeing boating and yachting as ways to distinguish Tampa Bay as a playground for the wealthy. Boating had already been a popular pastime around the bay for years. The Tampa Yacht Club, founded by local boaters in the first decade of the twentieth century, built a marina south of Ballast Point for members' use. In the 1920s, pleasure boating grew in popularity as combustion engines brought speedboats and yachting became a sign of conspicuous consumption. Boosters touted the sheltered waters of Tampa Bay as the perfect place to put in. "Little wind and tide make Tampa Bay ideal," proclaimed one.[37] Like Davis, developers planned yacht clubs or at least marinas for the buyers they hoped to attract. Developers billed Pinellas County's Shore Acres as "a Venice for water sports" because of its "unequaled waterfronts on bay, bayous, lakes, and connecting waterways."[38] At Beach Park, excavators dug out a sandy pine flatwood just off the bay to build a yacht club in 1925.[39] The Oldsmar Land Development Company built an "immense yacht basin" on its property and bragged its "ample facilities" had attracted at least one boatbuilder to relocate there.[40]

Yachting facilities revealed how the balance between public and private access to the bay evolved in these years. Private boating facilities created space for their members; others had to find public space suitable for launching or tying up. In Tampa, the lack of a municipal marina before 1927 meant that "many owners of fine pleasure craft . . . eliminated [the city] as a port of call in their annual tour of Florida waters." When one finally opened on Davis Islands that year, local boaters rejoiced, but within a few years it proved to be too small to meet demand.[41]

St. Petersburg's leaders took the opposite approach. In the aftermath of the 1921 hurricane, the city's waterfront underwent additional renovation, including the completion of Bayboro Harbor. City officials directed that the fill from the project be used to finish a municipal yacht basin along the city's waterfront. Dredged to ten feet and consisting of three basins divided by two large landings, it ran (as it still does) from 5th Avenue North to 5th Avenue South. A perfect addition to the waterfront park Straub had championed, the marina epitomized a commitment to public access. Slips could be rented for relatively low rates ($25–$75 per year, depending on size; about $450–$1,200 today), and anyone could dock for up to ten days at no charge. "This basin was built primarily for the benefit and use of tourists and residents, and it was not intended to be a revenue producing utility," the city's director of public works explained in 1923. "Thus while the basin is extensively used, much enjoyed and of great benefit to many people, the benefits cannot be measured by the financial yard stick."[42]

Figure 8. Municipal Beach, St. Petersburg, 1948. Tampa Bay has few natural beaches, but that did not stem demand for recreational access to the water. In 1913, St. Petersburg built Spa Beach as part of its original Municipal Pier project. When the city built the Million Dollar Pier in 1926, it expanded Spa Beach. In addition to the beach, the new pier had a public swimming pool and solarium. Until the end of segregation, Spa Beach was a whites-only beach; African Americans had to use a less developed beach at South Mole. While the pool and solarium closed in the 1960s, Spa Beach is still part of the St. Petersburg Pier complex and remains a popular spot to soak up the sun and splash in the water. Credit: Burgert Brothers Photographic Collection, courtesy of Tampa-Hillsborough County Public Library System.

The Million Dollar Pier soon complemented the marina. Led by *St. Petersburg Independent* editor Lew Brown, citizens raised money to construct the pier as a replacement for the storm-damaged Municipal Pier. Completed in 1926, the concrete pier shot more than a quarter mile into Tampa Bay. It included a solarium, a two-story casino with a ballroom, and an improved Spa Beach. A St. Petersburg landmark until it was demolished in 1967, there was no place like it anywhere else around the bay.[43]

Roadbuilding in this era eased movement around the bay and aided the growth of new areas. Before the 1910s, traveling was time-consuming and arduous. On the western side of the bay, the Tampa & Gulf Coast Railway carried riders between Tampa and Clearwater via Safety Harbor; those wishing to go to St. Petersburg had to change railroads. Steam-powered ferries cut out the hassle but saved little time. By boat or train, the trip usually took two hours each way.[44] Getting around the bay's eastern shore was more difficult. There was no rail connection until the Tampa Southern Railroad built a line on the east side of the bay in 1918. Automobile roads were scarce and rutted. Boats could be hired, but steamer service was irregular. In the 1910s, Bay Area advocates of the Good Roads Movement—a nationwide campaign to build roads and highways worthy of the era's new automobiles—pushed for new roads to connect the region. In 1910, county commissioners approved the Tampa Road around the northern side of Old Tampa Bay.[45] Rocky Point, Tampa Shores, Oldsmar, and Safety Harbor grew as a result. Later that decade, commissioners funded the building of the Bayshore Road to connect Tampa with Ruskin and Palmetto.[46] It soon became part of the Tamiami Trail, the first road to connect Tampa and Miami, and spurred the development of Gibsonton.

New bridges and causeways further sped trips around the bay by freeing passengers from ferry timetables. The first was the Gandy Bridge, designed to span the six miles of Old Tampa Bay between the Interbay Peninsula and St. Petersburg. George Gandy began plans for the bridge as early as 1910 and completed the first survey in 1918. Construction began in 1922 and the bridge opened in November 1924. Thirty thousand people and 7,500 cars crossed it during the opening day celebration. An engineering marvel for its time, the bridge complex consisted of two causeways extending from either shore (two and three-quarter miles on the Pinellas side and three-quarters of a mile on the Tampa side), each rising ten feet above the water. They connected a two-and-a-half-mile central concrete span supported by pilings that held it eleven to twenty feet over the bay. In the middle was a double jackknife steel drawbridge with seventy-five-foot openings. Despite costly tolls—roughly twelve dollars today—most thought it well worth it to shave twenty-four miles off the trip. By 1926, Gandy was already planning to widen the bridge.[47]

Real estate developers bought up land at the ends of the bridge hoping to catch drivers as they came across and entice them with waterfront property before they had a chance to look anywhere else. Developer Eugene Elliott eyed Weedon Island for his Florida Riviera project. It failed. But N. J. Upham and Perry Snell had better luck with Shore Acres and

Snell Isle, where they followed the pattern of dredging and filling they had used around Coffee Pot Bayou.[48] Snell Isle alone was built on 400,000 cubic yards of dredged bay bottom piled behind a ten-mile seawall.[49]

Tampa's 22nd Street Causeway across McKay Bay, completed in 1926, paled next to the Gandy Bridge, but developers saw it as no less of an opportunity. Developers hoped it would spur a real estate boom like the one Gandy unleashed in northern St. Petersburg. In 1925, a group of investors spent $4.5 million on a 2,400-acre tract that included more than six miles of bayfront property running south from the Palm River.[50] Calling it Tampa Beach, they envisioned something like George Merrick's Coral Gables. They boasted they would spend $100 million ($1.6 billion today) on improvements, including a six-and-a-half mile seawall, a beach, a yacht club, and artificial islands on which buyers could build homes.[51] On the day the causeway opened in early 1926, the developers brought in football hero Red Grange and his Chicago Bears teammates to woo potential buyers.[52] The timing was awful. Seven months after the causeway opened, the Florida land boom crashed when the 1926 hurricane tore through Miami and revealed its shaky economic foundations. With buyers scared off, Tampa Beach went bankrupt before the seawall went up.[53]

Despite the collapse of the real estate market, B. T. Davis (no relation to David P. Davis) decided to build a causeway across Old Tampa Bay from Rocky Point to Clearwater. Davis figured Tampa beachgoers would be willing to pay to cut at least twelve miles off the trip to the Gulf beaches.[54] Davis's dredge started digging up the sand from a shallow bayou at Rocky Point in February 1929. Pretty soon, workers pumped sand around the clock, dumping five cubic yards of shell-rich bay bottom at a time onto the more than eight-mile-long causeway.[55] Work slowed with the onset of the Great Depression, but Davis eventually secured a $600,000 loan from the federal government's Reconstruction Finance Administration to complete the project in 1934.[56] More than five years after he started, Davis began collecting his tolls.

Davis's delay and subsequent turn to the federal government was emblematic of the effects of the Great Depression on the Bay Area. Private development, which had slowed after 1926, ground to a halt after 1929. Tourism also faltered when the economy tanked. Most who came were looking for ways to avoid harsh winters on the streets of the nation's industrial cities, not for a round of golf or a guided fishing tour. The bay's reputation as a great place to catch one's dinner took on new meaning for thousands of Bay Area residents suffering through hard times. Franklin Roosevelt's response to the Depression shaped Florida and the nation after 1933. By the end of the decade, the New Deal's fingerprints could be seen around the bay. Funding for civil works projects served both to create employment opportunities and to build infrastructure to help places rebuild their economies. Since the Bay Area relied so much on tourism, several projects aimed particularly at improving the region's appeal.

The most obvious of these today is Tampa's Bayshore Boulevard. In September 1935, the federal government approved funds for the Works Progress Administration to build a new seawall "parallel to the old wall but 85 feet out in the bay." On the new filled-in shoreline, would be a "50-foot roadway," a "10-foot sidewalk," and "a space for beautification."[57] Designers came up with the famous balustrade in the design process. When completed in 1938, the Bayshore instantly became Tampa's calling card to the world. With its views of Hillsborough Bay and Davis Islands on one side and its fine houses on the other, it offered an aesthetic appeal not available along many other parts of Tampa's waterfront. Guidebooks advised tourists that taking a walk down the "world's longest sidewalk" was the thing to do.

The Bayshore project demonstrated a renewed commitment to public access on the waterfront, but private landowners shaped the project. The new seawall, by pushing the waterline back eighty feet, provided additional protection for the priciest houses in town in the event of storms. Meanwhile, the sidewalk and its famous balustrade provided little access to the water and no space for seaside recreation. Early plans for the Bayshore project included a small beach, but neighboring homeowners petitioned to have it removed from the plans. "We are opposed to the construction of any sand fill or beach in front of this beautiful driveway," they told city leaders. They feared it would be "unsightly and will afford a gathering place for evil-minded and disorderly persons."[58] It was almost inevitable that local residents soon complained that the benches installed along the balustrade were "littered with remnants, dead bait, fish entrails, and other debris repulsive to eye and nose" left behind by recreational fishermen. The benches, they fumed were ruined "as resting places for strollers desiring to contemplate the beauties of the Bay and its wild life."[59]

The view along the Bayshore was no substitute for access to the water. Many Tampans wanted a beach like St. Petersburg's municipal beach.

Spa Beach was wide, clean, and featured numerous amenities, including a freshwater pool and a solarium that permitted nude sunbathing. The city's recreation department held events to keep the crowds entertained. By the 1940s, beachgoers who did not wish to go all the way to the Gulf beaches piled in there.

But Tampa had no central waterfront to develop, and most other sites had serious problems. Dredging activities around Hooker's Point silted in the beach at DeSoto Park in the mid-1920s, so Ybor City and east Tampa residents decamped to a spit of sand at the base of the 22nd Street Causeway.[60] But it soon proved deadly. In the summer of 1927, at least seven people drowned in the offshore holes left by the dredging. Hillsborough County hired lifeguards and installed safety lines, and the small beach remained open for decades before high levels of pollution led to its closure.[61] The 1921 hurricane destroyed the swimming beach at Ballast Point. When locals sought to have it reopened, public health officials deemed the area too polluted.[62] Some advocated for a beach on Gadsden Point, at the tip of the Interbay Peninsula, where rustic outings and fishing trips had long been popular. In the mid-1930s, a privately operated beach for Black bathers operated there for a short time.[63] According to one report, beachgoers had access to a "bath house . . . a beer garden, grill room, tennis court, baseball diamond, dance pavilion, boating, fishing, and spring board."[64] But otherwise, Gadsden Point was remote, mosquito infested, and prone to rip currents, so planning never got far. The military's seizure of the property to build MacDill Army Air Field in 1940 foreclosed the option of a public beach there. By that time, though, all of Hillsborough Bay was out of the question because of bacterial contamination.[65]

The water was cleaner on Old Tampa Bay, but private developers controlled the most convenient parts of the waterfront. Picnic Island remained popular into the 1920s. In 1925, new owners branded it Port Tampa Beach. They hired jazz bands and offered "clean fresh salt water bathing, [a] dandy dance floor, [and] good snappy music."[66] It lasted a few years before petroleum shippers bought up the island for their storage tanks. Excursions to Sunset Beach remained popular until it closed in 1925. The beach turned out to be a sales gimmick, and the developer had no interest in maintaining it after the last lot sold. The developers offered the untended spot to the city in 1934, but state health officials declared it was too polluted.[67] People put their toes in the water at Frazier's Beach, located where

Grand Central Avenue (now Kennedy Boulevard) ended at the water, but it wasn't much of a sandy beach and was prone to mosquitos.

The Courtney Campbell Causeway soon became the hot spot for Tampa beachgoers. After purchasing the Davis Causeway in 1944 (renamed for State Road Board member Courtney Campbell), the state spent nearly a million dollars to plant invasive Australian pines and build concrete picnic facilities. Beachgoers soon overwhelmed the capacity. On weekends, "some are out there as early as 5:30 in the morning, seeking a picnic shelter overlooking Old Tampa Bay," a reporter noted in 1949. Those not lucky enough to get a spot "just park on any of the off-highway areas . . . and eat off the ground." These spaces lacked formal rules or oversight and were a bit anarchic. "The only admonition given carefree merrymakers by the State is shown on a sign printed in English and Spanish: 'This parkway for public use; please help keep it clean.'"[68] People fished, crabbed, waded, swam, sunbathed, and did pretty much as they pleased. Nighttime parties on these beaches were a rite of passage for many young people. Free, accessible, and largely unmonitored, the causeway was the domain of everyday folks.

The popularity of the Courtney Campbell Causeway drew the eyes of Tampa leaders to Rocky Point as an ideal spot to build a municipal beach. Largely undeveloped and glistening white from the sand dug up to build the causeway, the site became more appealing in 1946 when property owners offered to give land at the foot of the causeway for the city's beach.[69] While city leaders balked at the strings attached to the free property, they decided to build the beach on the south side of the causeway itself. In February 1950, Mayor Curtis Hixon unveiled a $1 million plan (nearly $12 million today) to build a wide beach, an Olympic-size swimming pool, an air-conditioned casino building with changing facilities and a restaurant, and parking for 1,000 cars. Initial plans even included an ice-skating rink! Supporters promised it would be "one of the finest beach recreation areas in the state."[70] One member of the citizens' commission, organized to advise the mayor about the beach, explained that the plan was to "provide something for the people who can't afford . . . beach clubs."[71]

Building the beach turned out to be far more complicated than simply planting an umbrella in the sand. The legislature had to pass special legislation authorizing the beach and cancel oil drilling leases approved for Old Tampa Bay. City leaders had to assuage Ybor City residents who preferred to see the money spent on improving the 22nd Street Causeway.[72] A. J. Simms and W. Howard Frankland, who offered the city a lease on a large waterfront property north of Frazier's Beach, argued that the roadside location would cost too much to dig out, take too much sand to build, and would create dangerous conditions offshore.[73] (Their property became a popular unofficial beach before it became the eastern base of the Howard Frankland Bridge in 1960.) Some worried the cost of the amenities would saddle the city with too many long-term obligations. Others doubted the value of spending money on things like cabanas, for which users would pay daily fees. "I thought the project was for the people who can't afford to go to the Gulf Beaches," said one citizen.[74]

The most ardent protests, however, came from Tampa's Black community leaders. They did not oppose the beach per se, just how it would be operated. The new municipal beach and its facilities would be open to whites only. City leaders had not even considered the desires of Black bathers in the planning. For a decade the city ignored the Black community's requests for a municipal pool for Black swimmers, so the snub added insult to injury. (The city already had several whites-only public pools.) C. Blythe Andrews, editor of the *Florida Sentinel*, and Perry Harvey, head of Tampa's longshoremen's union local, appeared before the city council and outlined the beach as one of a series of problems confronting Black Tampans. Their allies on the council agreed. One criticized the "fantastic plans" and argued that if "some of the fancy Dan things are removed, I think the Negro beach could be built also." Mayor Curtis Hixon stood by his plan, though, and promised a separate beach would be built "in the next couple of years in a location best suited to the needs of the Negro citizens."[75]

A decision by the Supreme Court that year made it difficult for Hixon to rely on this old delaying tactic. In its case ruling, the court required the City of Miami to provide a course for Black golfers as it had provided one for whites. Tampa city councilors voted to move forward with issuing the bonds to finance beach construction, but fears of a lawsuit led them to drag their feet on the once-popular project. For more than a decade, the city rolled out the plans whenever public sentiment for a beach grew but refused to move forward with building facilities. "The issue of segregation" is a "stumbling block in the path of a beach," explained one observer.[76]

The city eventually opened a municipal beach on the Courtney Campbell Causeway in 1961.[77] Other than sand and lifeguard stations, its only facility was a bathhouse built at a cost of just under $6,200.[78] Within

Figure 9. Ben T. Davis Beach, c. 1971. In 1950, the City of Tampa announced that it would build a municipal beach along the Courtney Campbell Causeway to satisfy demand for a swimming beach on Old Tampa Bay. Those seeking access to the water had been parking along the causeway for years. In addition to a safe beach, plans included a number of entertainment facilities. The city's plan failed to provide any access for African Americans, though. Fearing lawsuits, the city delayed its plans to build the beach for more than a decade. In 1963, Ben T. Davis Beach opened. Named for the visionary who built the causeway, its facilities were meager compared to those detailed in original plans for the site. Nevertheless, Tampans who did not wish to drive to the Gulf beaches found it a nice place to cool off. Credit: George S. "Skip" Gandy Photograph Collection, USF Libraries, Special Collections, Tampa.

weeks of its opening, white beachgoers complained about Black bathers using the beach, but the police refused to make them leave.[79] Official desegregation of Tampa's beaches had arrived. Tampa's white community had denied themselves much in delaying it.

In St. Petersburg, the same stumbling block tripped up political leaders' commitment to public access at Spa Beach. Since its creation, the beach and its facilities had been whites-only. Black St. Petersburg residents had for decades pointed out the inequities between Spa Beach and the beach at South Mole. Their voices grew louder as they watched the city devote more money to expand whites-only beaches on the Gulf. "Have any of you stopped to realize that our [sewage] waste disposal plant is less than one mile from the place which the city of St. Petersburg has allotted for our beach?" journalist Cal Adams admonished readers of the *St. Petersburg Times* in 1952. "The beach is scattered with bricks and weeds. Persons are using it as a fishing hole. Sea weed has washed all along the short shoreline and bottles are strewn all over the place," he explained. The city's failure to provide anything more than a rudimentary shack for a concession stand was made more ironic by its use of part of the mole as a storage area for "surplus bricks" and paving materials.[80] In early 1955, the city announced plans for a new beach for Black bathers near Boca Ciega Bay. Black leaders pointed out that the site was "muddy and improperly located."[81] Fearing that a neighborhood for Black homebuyers planned near the beach would drive down the value of their developments, white developers threatened lawsuits and killed the plan. Instead, the city promised to spend the money at South Mole.[82]

By this time, though, equality, not separate but equal, had become the goal of St. Petersburg's Black community. In August 1955, members of the Civic Coordinating Committee, a local civil rights organization, decided to test the strength of the 1954 *Brown v. Board of Education* decision and staged a "swim-in" at Spa Beach by attempting to purchase admission to the pool. They failed when the attendant refused them. The mere attempt riled many in the white community. Committed segregationists formed the St. Petersburg White Citizens' Council to urge the business community to resist integration. Rather than continuing to engage in nonviolent direct action at the beach and pool, the Civic Coordinating Committee sued the city for maintaining segregated facilities. They won. *Alsup v. St. Petersburg* took about a year and a half to wind through the courts, but in the end the city was ordered to give up its whites-only policy at its beaches.[83]

Black St. Petersburg residents waited until June 1958 to see if the city would obey the courts. On 5 June, eight teenagers and college students entered the beach and swam out into the bay. Unlike in other Florida cities, neither city officials nor white beachgoers overtly resisted them. Despite this, City Manager Ross Windom ordered the pool closed for three days. The day after it reopened, nineteen-year-old David Isom reintegrated the pool and swam for more than an hour. Windom closed the pool again and the city ordered it closed for the rest of the season, preferring to lose the revenue it generated rather than see it integrated. Instead of peaceful integration, city leaders pursued a course of resistance that both caved to and emboldened die-hard segregationists, some of whom responded to the integration of the beach and pool by leaving kerosene-soaked crosses at various points near the beach.[84]

With Spa Beach closed, St. Petersburg leaders attempted to sidestep the court's decision by setting up a section of the beach along the Gandy Bridge causeway for Black beachgoers. They promised to spend $16,000 fixing it up. City officials staged a hasty groundbreaking for the site four days after Isom's swim; they said it would be ready in two weeks.[85] In proposing this, they followed the example of Clearwater, which fought resistance from neighboring white landowners for more than a decade to build a beach for Black bathers on the west end of the Courtney Campbell Causeway to keep them off the city's Gulf beaches.[86] It opened the same week St. Petersburg proposed the Gandy causeway beach and turned out to be a dangerous spot with many holes and no lifeguards.[87] When it became clear that Black swimmers would continue to integrate Spa Beach, Windom pushed to have the pool closed and a cultural center put in its place. Most city councilors had no interest in pushing resistance that far and instead accepted that the beach and pool had to be integrated, which they were by 1959.[88] Their commitment to maintaining them as public resources wavered, though. While the beach remains, the city closed the pool by 1963.

When the Sunshine Skyway closed the gap at the mouth of the bay in 1954 and the Howard Frankland Bridge became the third span to cross it in the 1960s, they were symbolic of the growth taking place all around Tampa Bay. Between 1945 and 1980, the three counties around the bay grew four times over. Pinellas County, the state's second smallest in area, was the third largest in population by 1980 with more than three-quarters of a million residents, nearly six times more than in 1945. While much smaller, Manatee County witnessed about the same rate of growth in this era, although much of it was not on Tampa Bay. Hillsborough County, where the population more than tripled, paled only in comparison to its neighbors. All told, roughly ninety people moved to the Bay Area every day for those thirty-five years. And the census didn't count the thousands who lived there only part of the year or the millions of tourists who visited.

All these people in search of a Florida playground needed places to stay, so real estate developers started their bulldozers and got to work. Their techniques were not much different than those of earlier generations. Housing developments popped up seemingly everywhere, but waterfront property was understandably in greatest demand. Developers, it seemed,

Figure 10. Aerial view of Bay Crest, c. 1969. As demand for waterfront living grew in the middle of the twentieth century, developers brought dredge and fill to nearly every corner of the bay. Their work changed the shoreline forever. In this image, facing east from the north end of Old Tampa Bay, the white sands along the water mark the extent of filling to build the Bay Crest community. The sands had been pumped from the shallow seagrass flats to their south. Also visible in the upper right is Dana Shores, which had been built using the same method a decade earlier. Between the two is a remnant of the original shoreline. Much of it would be developed in the years after this picture was taken. Credit: George S. "Skip" Gandy Photograph Collection, USF Libraries, Special Collections, Tampa.

aimed to snatch up and develop any and all available shoreline, including much of it that was still under water.

The postwar brought an all-out assault on the bay's seagrass flats, salt marshes, and mangrove forests. From Bahama Shores to Venetian Shores on the west around to Bay Crest to Dana Shores on the north to Apollo Beach on the east, builders constructed bulkheads shaped like fingers from the shore, filled them with millions of cubic yards of cheap bay bottom, covered them with topsoil, and built tens of thousands of new homes, condominiums, and hotels on top. The fingers gave the developer more waterfront to sell. The canals, the exact obverse of the fingers next to them, allowed buyers to dock their boats in their backyards. It was the mid-century Florida dream.

Boca Ciega Bay, at the southern tip of the Pinellas Peninsula, became ground zero for the practice in the 1950s. Naturally shallow with a carpet of turtle grass, rimmed by mangroves, and located near the mouth of Tampa Bay, Boca Ciega Bay was one of the bay's most productive estuaries. Situated perfectly to mix the fresh and salt waters, it was a fishing paradise. Not pristine, its shoreline and bottom had not been significantly altered before the mid-twentieth century. That changed in the 1950s. "At least three dredges were in operation," a field scientist reported in 1953. "Many of the islands along the west shore were being enlarged, and the channels between the islands and beach mainland were being filled in. Although there were still some scattered stretches of mangrove, it appeared that these would soon succumb."[89] The dredging sparked resistance at the local and state levels and changes in state law regarding submerged lands (more on that later), but local politicians lacked the will to stand up to dredging interests.

In 1959, county commissioners permitted the most egregious of the dredging plans presented to it—the so-called Ratner Fill—that sought to build up 515 acres that would block much of the bay's natural flow. Court fights delayed the start, but by 1961 the project was completed. (Drivers on the Pinellas Bayway today cross it when going through Bayway Isles and Isla Del Sol.) The ecological effects of the fill were obvious quickly. "By 1964, this once-shallow coastal lagoon, with its vast meadows of turtle grass, had been transformed into a channelized cesspool," historian Bruce Stephenson has explained. The clear water became turbid and polluted, the sandy bottom replaced by gooey muck caused by effluent dumping and runoff. By the end of the decade, a quarter of the bay had been filled, and it had become the state's most polluted bay.[90]

Boca Ciega Bay made it clear to those that had seen it before dredging that their playground was on fragile ground. Anyone living around the bay in those years could have sensed the same wherever they looked. The very things that drew people to the bay—its aesthetic charm, its bountiful fisheries, its warm water and salt air—came under assault as people moved recklessly, both as individuals and as polities, to grab their place by it. All the building, all the digging, all the dumping caught up with them faster than they could imagine. By the middle of the twentieth century, the bay was a sewer, and everyone knew it.

5

SEWER

In the 1970s, having journalist Mike Wallace at your door was a sure sign something was wrong. The veteran coanchor of the CBS news program *60 Minutes* specialized in digging up corruption. He seemed to enjoy making the powerful uncomfortable. He was blunt and backed down from no one. And there he was in Tampa in late 1973, interviewing county officials, bankers, developers, and residents. What wrong was he pursuing? Unregulated growth.

The interviews aired on 10 February 1974, in a seventeen-minute segment called "Bigger Is Better?" In the report, Wallace narrated a litany of problems—water shortages, habitat loss, pollution, and sprawl, among others—over a montage of skyscrapers, suburban homes, and raw sewage. Looking to explain the causes, he interviewed Fred Church, president of Tampa's First Federal Savings and Loan, whom he left flummoxed with questions about the consequences of growth. Riding through Hillsborough County's Town 'n' Country neighborhood in developer James Shimberg's convertible, Wallace asked Shimberg why development went on unabated when it created so many problems. Shimberg (who, it should be said, was civic minded and sensitive to the problems of growth) pointed to the failure of political leaders to establish guidelines. When asked to respond to Shimberg's assessment, Hillsborough County Commissioner Carl Carpenter pivoted to a traditional pro-growth stance. "We don't feel that we're in the position, as elected officials, to have a plan that says, 'This is it. These are the lines, and this is it, and under no circumstances will it change,'" he explained. "We feel that we need a flexible plan if we are going to hold on to our basic form of government." Asked whether rumors of bribery greased the wheels when it came to development, County Commissioner Betty Castor, a former teacher whose political career started with environmental activism, paused, stifled a smile, and responded, "Well, no, I don't think that that goes on here. But I believe that the public thinks that goes on here."[1]

Hillsborough County's Director of Pollution Control Roger Stewart was less cagey when Wallace interviewed him. Riding in a boat on the fingerlike canals of the Baycrest subdivision on the north side of Old Tampa Bay, Stewart explained the situation. "This is a residential canal, Mike, but the water quality is such that I guess you call it an open sewer." "This community," he continued, "is served by a sewage treatment plant that serves only this community . . . and when these people flush their toilets, it goes through the plant and is discharged into their very backyards." Is the water is polluted? Wallace asked. "Environmentally, or from a pollution standpoint, it is totally polluted." Wallace pushed further. Did the builders or buyers know where the sewage would go when they built the houses? "I frankly don't think anyone gave a damn at that point," Stewart replied.[2]

The exchange was classic Stewart. Since being hired by Hillsborough County Commissioners and the County Health Department as the first Director of Pollution Control in 1970, he had never hesitated to speak bluntly about the problems he saw. His first year on the job, he called for a "total restriction" on dredge-and-fill operations and criticized plans for a $63 million sewage treatment plant that he said would be ineffective and leave the bay as "a commonly owned cesspool."[3] Threats from developers and politicians did nothing to quiet him. When asked by Wallace about Stewart's claim that the county commission did little to deal with pollution, County Commissioner Carpenter visibly swallowed his disdain and replied simply, "Some of those people, of course, we're not real interested in pleasing."[4] The feeling, it might be said, was mutual.

Try as some might, no one could deny Roger Stewart's assessments of the pollution in Tampa Bay. An Air Force pilot turned marine scientist, Stewart was in the best position to know since he had worked as part of an eleven-person team that studied pollution in Hillsborough Bay. In 1966, Representative Sam Gibbons invited the Federal Water Pollution Control Agency (FWPC) to come to Tampa hoping it could solve a problem that had vexed residents of Bayshore and Davis Islands for decades: noxious sulfur fumes emanating from Hillsborough Bay that were bad enough to make people sick and powerful enough to corrode metal.[5] Congress created the FWPC in 1965 to assist cities in solving their water pollution problems, and Gibbons seemed to think the agency would offer a quick fix to remove the rotting algae that caused the stench. Instead, the agency pursued a deep study of the problem. The team worked more than a year taking "samples of water from the bay at all hours of the days and nights— and in all kinds of weather." By the end, "they probably knew more about the water in the bay than anyone has ever known."[6]

The team's report—entitled *Problems and Management of Water Quality in Hillsborough Bay, Florida*, but better known as the Hagen Report after the team's leader, John E. Hagen III—was the first comprehensive study of the pollution entering the bay and its effect on its health.[7] In it, the authors weaved together scientific data, anecdotes, and photographs to demonstrate what anyone who cared to look knew: Hillsborough Bay was nearly dead, the victim of a thousand cuts of varying sizes and lethality. While its findings applied only to Hillsborough Bay, the study seemed prophetic about what fate lay in store for every corner of Tampa Bay. For Bay Area residents, the Hagen Report offered a model for beginning to understand what was going on: political leaders, industrialists, developers, and, by extension, everyone, had treated Tampa Bay like a common sewer.

The threats to the bay's health were interconnected and they compounded each other. Trying to explain them was like trying to explain a knot made up of multiple strings. To detangle the knot, scientists generally divided the problems into six overlapping categories: pollution; eutrophication (severe reduction in dissolved oxygen in water); circulation; turbidity (a measure of how much light penetrates the water); habitat loss; and biotic decline, the loss of sea life, both plants and animals.

Dividing up the problems made it easier to understand how each problem affected the others. Take the stench along Bayshore Boulevard that Gibbons had hoped to fix. Sewage runoff dumped nitrogen and phosphate plants dumped phosphorus into the water. Weak tidal flow allowed these to concentrate in Hillsborough Bay. Algae on the bay bottom, in this case *Gracilaria*, proliferated because of the nitrogen and phosphorus. The algae died when low tides exposed it to the air. Siltation along the western shore of Hillsborough Bay, caused by redirection of water discharged by the Hillsborough River after the construction of Davis Islands, left more *Gracilaria* exposed at low tide. Decay of the algae released sulfur into the air, which made the air around the shoreline unhealthy.[8] The smell was an ecological problem, of course, but economics, culture, and history shaped

Figure 11. Polluted water warning, MacDill Air Force Base, c. 1966. Scientists from the Federal Water Pollution Control Administration descended on Tampa Bay in the mid-1960s to determine what could be done about the odor generated by dead algae in Hillsborough Bay. The team spent more than a year studying the entire bay. Its conclusions, published in what came to be known as the Hagen Report, documented that Tampa Bay was under dire threat from pollution. Credit: Federal Water Pollution Control Administration, *Problems and Management of Water Quality in Hillsborough Bay, Florida* (Washington, D.C.: Government Printing Office, 1969).

the Bayshore stench. The smell itself also presented a social problem that residents turned to political leaders to solve. Detangling that political knot proved far more difficult than describing the scientific one.

A stinky shoreline can result from many causes. In August 1915, Tampans endured a three-day stench emanating from millions of rotting fish, crabs, and all manner of dead sea life that clogged Hillsborough Bay from downtown to Catfish Point.[9] People all over town covered their noses when they ventured outside. Housewives perfumed their curtains to block the smell from their homes. The chemicals released by the rotting fish caused metal to oxidize and paint to peel off houses. No one could explain the fish kill. Some blamed the activities of phosphate companies along the Alafia River. Old fishermen who said they had seen something like it in the decades before the phosphate mines were built rejected the idea.

Stagnant rainwater blown in from wetlands east of the bay caused it, they said. Others described a sheen on the water, and some asked if leaking oil from a ship may have had something to do with it. Few believed this to be the cause but agreed that the water looked different and smelled foul in the days before the die off. Descriptions of the fish kill point to a rapid eutrophication event, a sudden reduction in the dissolved oxygen in the water often caused by algal blooms. Algal blooms can arise naturally, as can other phenomena that can cause rapid eutrophication, so from this distance in time, it is impossible to know if human activities caused or contributed to the fish kill. What we can tell, though, is that by the early twentieth century Bay Area residents already had a sense their expanding cities, factories, and ports were polluting the bay.

The pollution would have been hard not to notice. Before 1914, the city of Tampa dumped sewage directly into the bay. "The odor from foul sewage salute[s] the visitor immediately upon his arrival at Tampa's gates," one local noted. "The sewer that dumps its offal near the Mallory Line wharf . . . is surely an offensive feature."[10]

Tampa's first plant, opened in 1914, simply held solids in a series of tanks to decay while releasing the liquid sewage into the Estuary east of downtown.[11] St. Petersburg installed a similar septic tank system on that city's southern waterfront in 1921. It, too, drained filtered effluent water directly into the bay. Engineers promised the filtered water would be 99 percent pure. It wasn't.[12] A growing population overwhelmed both systems. Hillsborough and Pinellas Counties nearly tripled in size between 1910 and 1930, not counting the increased number of tourists. In 1927, St. Petersburg found it necessary to replace the system it had built only six years earlier. By that time, Tampa's sewage treatment plant could handle only two-thirds of the waste it received.

Sewage also flowed unabated into the bay from other sources. By the 1920s, at least eighteen sewer lines dumped unfiltered effluent directly into Tampa Bay on St. Petersburg's north side.[13] On the Hillsborough side, the problem of unregulated sources was worse. The Hillsborough River, which collected runoff along its course, discharged millions of gallons of untreated sewage daily. Sewage from Davis Islands ran unfiltered through a pipe on the southern end. And development along the Interbay Peninsula dumped effluent into both Hillsborough Bay and Old Tampa Bay. "No chemical analysis is needed to show the condition of the waters of the bay," Tampa's health director asserted in 1927. "If one is unfortunate enough to go along Bayshore drive . . . when there is a heavy east wind, the polluted character of the water is only too evident."[14]

By 1927, the lower end of the Hillsborough River, McKay Bay, and Hillsborough Bay were off limits to swimmers. In 1928, Tampa officials threw up their hands and declared there was little hope of saving Hillsborough Bay from sewage. "I do not believe we will ever be able to clean up the water to any appreciable degree," Tampa's city engineer told a reporter.[15] Later that year, the developers of Gibsonton, south and east of Tampa on Hillsborough Bay's eastern shore, petitioned to have Tampa's dumping shut off. The pollution, they said, was "practically ruining [the town] from the standpoint of beauty, fishing, and bathing. The bay should be preserved for the benefit of citizens and visitors."[16] Their plea went unanswered. In 1931, holders of oyster leases near Gadsden Point sued to enjoin Tampa after the health department closed their bars to harvesting. They asked the court to force the city to stop dumping unfiltered sewage into Hillsborough Bay.[17] They lost when the Florida State Supreme Court decided that the city had a common law right to use the bay as a sewer.[18]

In an era with few records kept about sewage, the health of the oysters served as a pretty good measure of the extent of pollution. Oysters take up pollutants as they filter the water for nutrients. Surveys of oyster beds beginning in the 1920s showed things were very bad in some places and getting worse all around. In 1925, health officials in St. Petersburg warned consumers that "insanitary refuse" had made oysters taken near the city "dangerous to health."[19] The following year, Tampa health officials reported eighty-five people contracted typhoid fever from infected oysters; twenty-five died.[20] By 1929, Tampa's health department employed a warden to prevent commercial oyster fishers from harvesting any from McKay Bay down to Ballast Point. They eventually closed them to recreational catches, too.[21] By 1940, the State Board of Health closed "much of the water of Tampa Bay adjacent to St. Petersburg" to oystering "on account of the excessive sewage pollution."[22] In 1942, state officials barred oystering in the Manatee River because of pollution from Bradenton.[23] In southern Pinellas, Long and Cross Bayous were closed due to pollution by the early 1950s; Boca Ciega Bay followed in the 1960s.[24] Pollution was lower in Papys Bayou and north into Old Tampa Bay well into the 1960s, but by 1972, only ten acres of natural oyster reefs were open to public harvesting. All told, only 2 percent of Tampa Bay's oyster beds were open to harvest, and those were threatened.[25]

The post–World War II building boom around the bay accelerated the pollution crisis. The inadequacy of Tampa's system was obvious by the 1930s, but the Great Depression made expanding it out of the question. Strapped for cash, Tampa political leaders sought federal funds to have a new sewer plant built in the late 1930s, but the plans never came together. During the war years, when Tampa grew rapidly, many neighborhood sewage lines were simply hooked up to the storm drains. "Palma Ceians desecrated one of God's beauty spots" by "dumping raw sewage into the bay," said a resident of an adjacent neighborhood. Others complained that "nauseating odors" on both sides of the Interbay Peninsula were driving down property values.[26] It took until 1959 for Tampa to build a modern plant on Hooker's Point.[27] Within a decade, it was overwhelmed. By 1968, the Tampa treatment plant was removing only 60 percent of solid waste from the 26 million gallons of sewage it received each day; the plant pumped the remaining 40 percent directly into the channel south of Hooker's Point.[28] "I . . . seen turds floating in . . . the water," remembered one crabber.[29]

St. Petersburg's system proved adequate through the 1930s but reached capacity by the early 1940s and failed to provide the level of purification it promised.[30] St. Petersburg officials went into full-on panic when pollution threatened to close Spa Beach in 1949.[31] "I would think the city fathers and the Chamber of Commerce would want this beach cleaned up and maintained," wrote one woman who complained of having to "hear tourists . . . pass slurring remarks about the sloppy way Spa Beach is maintained."[32] (Few among the city's white residents and political leadership worried about the much higher levels of pollution at South Mole.) The city completed building an expanded plant to handle the problem in 1954, but, like Tampa's, it still sent treated effluent into the bay.[33]

Fixing the sewers of Tampa and St. Petersburg would not have solved the problem, though, because they were not the only source of effluent pollution. Effluent flowed from private sewage plants as well. Most faced the same capacity problems that confronted municipal operators. In 1953, the U.S. government installed the first sewage system on MacDill Air Force Base.[34] Expansion of the base in the 1960s overwhelmed it, and, by 1969, the base was among the worst polluters of Tampa Bay.[35] A similar scene played out in the Bay Crest neighborhood, where Wallace toured with Stewart. Developers built a sewage plant for the neighborhood, but it was soon a blight. Not only did it exude an overwhelming odor at times, it also dumped lightly treated waste into the neighborhood's main access canal that led right to the waters north of the Courtney Campbell Causeway.[36] The Bay Crest plant was just one of at least eighteen dumping sewage into Old Tampa Bay in that area.[37] By 1970, it and most of the others had been cited for exceeding pollution levels.[38] In 1972, Hillsborough County's Environmental Protection Commission noted there were fifteen hotspots of coliform bacteria generated by sewage releases on its side of the bay. On the northeast side of the Courtney Campbell Causeway, scientists recorded "counts as high as 240,000/100 ml." For comparison, the Environmental Protection Agency considered any waters with more than 1,000 organisms per 100 ml to be unsafe for fishing or recreation.[39]

Off the utilities grid, many new houses, trailer parks, and even hotels relied on septic tanks that eventually flowed into the bay. For developers, septic tanks seemed to be the perfect solution to household wastewater. The tanks were cheap and easy to install, and the sandy soil of the area seemed to be an ideal environment for them. They were especially popular in rural areas, but builders in higher-density areas used them as well. As late as 1970, some neighborhoods on the Interbay Peninsula still relied on them.[40] But the tanks proved to be a nightmare. If installed incorrectly, they leaked. If installed over clay, they backed up. And even properly operating tanks could be overwhelmed by the Bay Area's drenching summer rains and allow untreated waste to escape. That effluent then flowed into creeks, streams, and canals that emptied into Tampa Bay. In 1952, Pinellas County officials blamed leaky septic tanks in Gulfport for rising pollution levels in Boca Ciega Bay.[41] Five years later, a report from the Greater Tampa Chamber of Commerce found that overflowing septic tanks caused high levels of the microorganisms that caused typhoid and dysentery in the Palm River and McKay Bay. (It didn't help matters that at least three meatpackers dumped refuse directly into the river as well.)[42] Residents in the vicinity of Rocky Creek, which flows into Old Tampa Bay, noted the effects of septic tank drainage. The creek had once been "clear with a white bottom," one longtime resident complained in 1968, but now "the water is filthy and not fit to swim in."[43]

Sewage was a problem of human health and aesthetics. It was also a source of large quantities of nitrogen. On its own, the nitrogen was bad enough for water quality. In Tampa Bay, it came together with another pollutant to make things even worse: phosphorus. Together they fueled algae growth on a scale that nearly killed Hillsborough Bay.

Phosphorus came from two sources. Some came from washing machines all over the Bay Area. In the 1950s, synthetic detergents replaced soap in many cleaning applications, especially laundry. By 1958, more than 90 percent of packaged household cleaning products were detergents, all of which included one or more phosphorus compounds.[44] Suds were the selling point for detergents and they soon showed up in bodies of water across the nation. Tampa Bay's size meant there were few bubbles, but the phosphorus from cleaning products was there. In 1969, Tampa's sewage plant alone dumped 7,300 pounds of phosphorus per day into Hillsborough Bay; other sources combined delivered roughly the same amount.[45]

Most of the phosphorus in Tampa Bay, however, came directly from the source: the phosphate processing plants that fed the port's growth. Miners had dug phosphate out of Bone Valley east of Tampa since the 1880s and the industry expanded in the early twentieth century. In the years following World War II, demand for phosphorus exploded as commercial fertilizers underwrote the growth in agricultural production worldwide. Between 1945 and 1970, Florida phosphate production increased by nearly 600 percent to 28.5 million tons.[46] By 1978, it rose again to more than 43 million tons.[47] Tampa's port boomed alongside the industry, but the industry was also killing Hillsborough Bay.

The process for turning phosphate ore (phosphorite) into phosphoric acid—the middle P in fertilizer manufacturers' N/P/K formula—increased the environmental consequences of phosphate mining. In short, it involved a lot of water. Once strip-mined out of the earth, phosphate ore is delivered to processing plants in the form of a water slurry. Mechanical processes then isolate the phosphorite from the clay and sand (called the matrix) carried in the slurry. Once isolated, the phosphorite is treated with sulfuric acid to create phosphoric acid. This leaves calcium sulfate (phosphogypsum) as a by-product. For every ton of phosphoric acid produced, five tons of phosphogypsum are created. And because it concentrates naturally occurring uranium and thorium (and related isotopes), phosphogypsum is radioactive and cannot be used in commercial applications. Instead, it is stacked in massive piles usually just called gypsum stacks. (Some today are greater than five hundred feet tall and cover hundreds of acres.)[48]

Handling what is called process water is even more complicated than dealing with gypsum. Because it dissolves chemicals and forms a thick clay slime in the manufacturing process, it cannot be returned directly to the watershed without causing immediate and significant consequences. Companies built ponds to hold this wastewater, sometimes at the tops of gypsum stacks. With little guidance to regulate their construction before the 1970s and a very rainy environment, they often leaked and sometimes failed entirely. In 1952, for example, a reservoir at the American Agricultural Chemical Company's plant near the Hillsborough County community of Boyette "collapsed to spill waste matter" into Fishhawk Creek. The creek then carried the waste into the Alafia River, where it left a "four inch deposit of waste materials along the shorelines." Owners of the riverfront properties sued, claiming the company was negligent in its construction and maintenance of the pond."[49]

This was hardly the only spill the Alafia absorbed. Once renowned for its fishing and beauty, the river was nearly dead by the 1950s because of the volume of phosphate and other chemicals reaching the river. Only about fifty miles long, the river nevertheless drains a wide watershed in the heart of the phosphate mining district. Polluted water from leaky ponds and overflowing reservoirs in western Polk and eastern Hillsborough Counties ultimately reached the Alafia. "There is so much evidence of chemical pollution in the Alafia that underbrush at the water's edge has turned white," one resident wrote in 1954. "Many dead fish float by, testifying to the wasteful and destructive practices."[50] In 1955, residents along the river pressured the Florida Legislature to pass the first law to protect the Alafia from spills, but it capped penalties at $500 per offense and left enforcement at the discretion of county solicitors. It did nothing to force companies to clean up the river.[51]

Despite the legislation, the problem grew only worse for the Alafia River, which meant the problem only grew worse for Hillsborough Bay. In the 1960s, fourteen processing plants operated near the headwaters of the Alafia River in Polk County.[52] Other plants along the river, including the Gardinier (formerly American Cyanamid) plant at the shore of Tampa Bay, added to the volume of runoff. By the late 1960s, the Alafia delivered more than 400,000 pounds of orthophosphates, which translated to 43,000 pounds of pure phosphorus, to the bay every day. The Alafia alone dumped 94 percent of the phosphate in the water.[53] "Throughout most of its length," a journalist put it in 1970, "the Alafia acts as a sewer pipe for the chemical industry."[54] (The river was also a significant source of nitrogen since it also collected runoff from citrus groves and ranches as it flowed.)

The combination of nitrogen and phosphorus created the perfect environment for algae and phytoplankton to bloom beyond normal bounds. The 1969 Hagen Report relied on comparisons of Hillsborough Bay and Old Tampa Bay to demonstrate how bad things had gotten in the former. Hillsborough Bay, the authors reported, had a phytoplankton population four times that in Old Tampa Bay. *Gracilaria*, the benthic algae that caused the odor along the Bayshore, thrived on the polluted water. It constituted 98 percent of Hillsborough Bay's attached algae growth, but only 2 percent in Old Tampa Bay.[55]

The phytoplankton and algae damaged the Hillsborough Bay ecosystem in several ways. Eutrophication, a severe decline in dissolved oxygen, was the first. As phytoplankton and algae died, they released carbon dioxide and other compounds into the water. *Gracilaria* is sensitive to changes in salinity, so freshwater runoff from both rivers and stormwater systems caused it to die back often. Meanwhile, the bacteria that consumed the dead algae proliferated, using up much of the dissolved oxygen in the water and thereby causing long-term depletion of fish stocks. Hillsborough Bay was not a dead zone, but, scientists and anglers agreed, fish were harder to find. "Mackerel," one sportswriter put it, "just don't show up in dirty old Hillsborough Bay too often." "Bottom scavenging" tarpon, he contrasted, continued to thrive in the waters. Fishermen in the annual Tampa Tarpon contest pulled them in from all over Hillsborough Bay, including right off Hooker's Point, but their presence belied how sick the bay was.[56]

Swimming along the bottom, a tarpon would have beheld with its giant eyes the other problem created by the overabundance of sewage and algae: a slimy bay bottom. The slow anaerobic decay of dead algae, the accumulation of untreated waste solids, particularly south of Hooker's Point, and weak tidal scouring action exacerbated by the construction of deep shipping channels combined to coat the bottom of Hillsborough Bay in a carpet of slime. The once-beautiful "white sand bottom," a crabber explained, "now is covered in 1 to 6 feet of mud."[57]

The mud, in addition to being ugly and, when exposed to air, smelly, caused two broader problems. First, it smothered bay-bottom ecosystems. Few other benthic species—blue crabs, for example, and other invertebrates—could long survive on the pudding-like slime that buried the natural sandy bottom. One 1960s study of Hillsborough Bay found no mollusks in the areas of the greatest concentrations of *Gracilaria* and an overall reduction in the number of species from twenty to four.[58] And mud, second, increased the water's turbidity, making it difficult for light to pass through. It did not take much, just the dragging of a net or the passing of a boat propeller to stir up the slimy mess, which tended to remain suspended in the water. "You can follow a freighter ten miles," aerial photographer George S. "Skip" Gandy recalled, "just follow the mud trail and you'll know where he is."[59]

Slimy mud alone did not make the bay darker, though. Phytoplankton blooms and suspended sediment from near-constant dredging operations contributed as well. "The turbidity was incredible," one scientist recalled. "You could stick your hand down and within ten inches your hand would disappear."[60] Turbidity was both an aesthetic and ecological problem. Murky water never gave anyone joy. And darkness made it impossible for seagrasses to grow even in areas where the bay bottom remained sandy.

The physical transformation of the bay and its shoreline compounded these problems both by altering water flow and destroying habitats. The worst consequences of this could be seen in McKay Bay. The construction of the 22nd Street Causeway and the expansion of Hooker's Point from the 1920s to the 1940s had already constrained the flow of water into and out of the bay, reducing tidal scouring, concentrating pollution, and fostering siltation. The construction of the Port Sutton and East Bay Terminals in the 1960s made the problem worse. During that same decade, as part of a larger package of flood control measures on the Hillsborough River, the straightening of the Palm River to make the Tampa Bypass Canal shot polluted freshwater runoff right into the bay.[61]

On Old Tampa Bay, the Gandy, the Howard Frankland, and the Courtney Campbell created similar flow problems. The causeway sections of each functioned like groins or dams in the water, blocking the flow and allowing silt to accumulate behind them. While the Gandy and Howard Frankland created their own problems, the Courtney Campbell (née Davis) Causeway, with its small bridges, was the worst in terms of Old Tampa Bay's health because it altered the flow of water into the large Upper Tampa Bay estuary.

Benjamin T. Davis had planned for his nine-mile causeway to have four bridges, but costs limited the final project to two, neither of which was close to shore; the western bridge had a tidal prism of only three hundred feet. With tidal flow reduced, silt accumulated on the north side. In 1948, only fourteen years after the causeway's completion, residents of Safety Harbor and Oldsmar requested that the Army Corps of Engineers dredge a channel through Old Tampa Bay to allow them to get their boats through Upper Tampa Bay. Resident after resident testified at public hearings about the mud accumulating along the shoreline and changes in water clarity. The sandy bottom around Safety Harbor's pier, one longtime resident explained to the engineers, was now "filled up with mud and silt." Another complained that the "water is getting so bad out there" that "no fish could live in here." According to one woman, the local Red Cross "did not consider the bay . . . a fit place for swimming lessons."[62] Two years later, after the Army Corps of Engineers refused to dig the channel, Safety Harbor officials again pleaded for relief. Shoaling behind the causeway, they explained, had "increased the tidal flats by hundreds of acres and mud has been deposited along the entire length destroying oyster beds and increasing a menace to the health and welfare of the public. These flats are . . . breeding places for mosquitoes which did not exist before the tidal flow was restricted."[63] In subsequent years, dredge-and-fill projects along the Courtney Campbell, both for beach creation and other construction, made flow problems worse.

Causeways were bad, but the problems created by dredge-and-fill projects were on a different scale in terms of destructiveness. Not only did they add to the pollution by putting people and their waste close to the bay, but their very construction came at the cost of irreplaceable habitats. Fill projects ripped up shallow seagrass habitats, dumped them on top of tidal marshes and mangrove forests, and altered the flow of water along the shoreline. Dredging and filling in Boca Ciega Bay in the 1950s had immediate ramifications for the bay bottom. "Canals between residential 'fingers' less than 10 years old now have sediments consisting of 70 percent silt, whereas these bottoms originally were 80 to 90 percent sand," a scientist reported in 1966. "These fine sediments are heavily charged with hydrogen sulfide and are devoid of invertebrates."[64] While some at the time hoped Boca Ciega Bay's example would serve as a cautionary tale, it instead was a harbinger of things to come. A litany of new projects followed in the 1950s and 1960s: Apollo Beach, Bay Crest, Coquina Key, Bahama Shores, Dana Shores, and Bay Port, to name a few.

The combination of pollution and construction destroyed most of the bay's seagrasses. "In Hillsborough Bay, most seagrasses have been killed,"

Figure 12. Siltation in Upper Tampa Bay, west end of Courtney Campbell Causeway, c. 1970. The original design of the Courtney Campbell Causeway severely restricted tidal flow and caused siltation on its north side. Within a decade of the causeway's construction, residents complained of not being able to swim or use their boats in the bay. In this image, the line of mangroves marks the historical shoreline directly north of the causeway, which appears as a long white band in the center foreground. Siltation also shallowed the water between Safety Harbor and Booth Point (*upper right*). Credit: George S. "Skip" Gandy Photograph Collection, USF Libraries, Special Collections, Tampa.

scientists noted in 1975, "and in Upper Tampa Bay, they are restricted to shallow areas near shore."[65] In the 1980s, ecologists estimated they covered an area only one-fifth the size they had at their last natural maximum.[66] Five different kinds of grasses grew natively throughout the bay: turtle grass; manatee grass; shoal grass; widgeon grass; and star grass. Turtle and shoal grass were the most common, star grass, the least. Meadows were monospecific, meaning each species had its own meadows, determined by a range of environmental parameters. As their names suggest, some were noted most by the vertebrate species that dined on or otherwise lived in them. (Manatees and green sea turtles will eat all kinds of seagrasses. They just happen to have favorites.) The loss of seagrasses led to systemic problems. Without their photosynthetic exchange of oxygen and carbon dioxide, the aquatic oxygen cycle grew worse. At the same time, the loss of grasses meant a loss of food for a wide range of vertebrate and invertebrate animals. It also meant the loss of protective habitat for many animals. By the 1960s, scallops had almost entirely disappeared from the bay. In the 1970s, manatees, whose numbers declined statewide in the twentieth century due to pollution and boat strikes, still made their way to the Little Manatee River during the winter, but the rest of the bay was a no-go zone for them because of the lack of seagrasses.[67]

The reduction of Tampa Bay's mangrove forests and tidal marshes by development projects rivaled the loss of seagrass meadows in importance. While no studies have pinned down the extent of each habitat's loss over the twentieth century, it is probably safe to assume that that most, if not all, development along the shoreline replaced one or the other. "Destruction of much of the natural mangrove and the accompanying free flow of upland sediment," a scientist noted in 1953, had led to the "formation of extensive tidal flats and the sedimentation of formerly productive bottoms."[68] By 1972, scientists estimated that only 1,728 acres of tidal marshes and 17,474 acres of mangroves were left; roughly half of each were in the lower, less developed parts of the bay.[69] Without mangroves to hold coastal soils in place, erosion from both tidal action, storms, and human landscape alterations increased erosion and siltation. The loss of those habitats also severely reduced Tampa Bay's productivity as an estuary as dozens of bird species, fish, and other aquatic species depended on the shelter and sustenance they provided. The loss of habitat directly affected the bay's fisheries, as catches of everything from mullet to shrimp declined.

Tampa Bay had reached a critical point in its history. Its health was clearly in danger. Government agencies and local people signaled the warning, as did the media. Many cast about for solutions. Others, however, seemed content to keep pressing forward on the path that had led to this point. Sure, they might agree that pollution should be limited, but they saw no reason to stop turning the bay into what they thought it should be. Planners at the Tampa Port Authority moved to deepen the channels further and expand the port facilities. Developers drew up plans for new, bigger, costlier dredge-and-fill projects. Politicians batted around ideas for closing off the Courtney Campbell Causeway entirely and turning Upper Tampa Bay into a freshwater lake and for building a massive hurricane gate from Interbay to Piney Point. Others talked of a fifth span across the bay.

Even in the wake of the *60 Minutes* exposé, it seemed pro-development forces might still win the day. Hillsborough County Commissioners fired Roger Stewart about a month after he appeared on the show. They claimed the firing had nothing to do with the appearance or his blunt assessments of their failings. Instead, they said they fired him for other comments, in another venue, about the pollution produced by Tampa's water treatment plant. He spoke out of turn at that meeting, they told him, so they voted three to two to fire him. It was the last of a long line of insubordinate statements, they said, and they could no longer have him "demoralize" the agency for which he worked by airing the county's dirty laundry on national television and insulting his bosses in the process. A few developers expressed their delight that the man who'd stood in their way would be out of a job, sure they would be able to replace him with someone friendlier to their demands. They might have gotten at least some of what they wanted, too, if the political winds hadn't shifted.

6

CAUSE

St. Petersburg hired Emil Hicks as Pollution Control Director in 1971 to help the city clean up its problems. Like Roger Stewart, his counterpart on the other side of the bay, this former trucking company manager from Detroit was not one to mince words. In a September 1973 water quality report, he concluded that "Obviously, humans should not be permitted to swim in all the waters of Boca Ciega Bay" and called for a moratorium on all construction until the city could develop a plan to manage it.[1] A week later, fearing his comments were bad for business, his bosses forced him to retract the report.[2] In a public meeting a week later, Hicks retracted his retraction.[3] Like Stewart, his bluntness got him fired. In fact, he was fired the very same month. City Manager Raymond Harbaugh accused Hicks of hurting morale when he "misled the public" in a report about water loss in the city's water system.[4] Observers charged that the accusation was, as in the Stewart case, a lame excuse to oust a vocal critic of growth and the city's failure to manage it. "Hicks' main problem was candor—he told it straight," said the editor of the *St. Petersburg Times*.[5]

Developers cheered, but Hicks's firing roused public anger. It was "an outrage," wrote Carol Andren, president of Suncoast Active Volunteers for Ecology (SAVE).[6] St. Petersburg resident Ruth Fullington agreed. "I am in a state of rage and utter dismay," she wrote, adding that it "makes it only too plain that our city administration wants a 'patsy' who will go along with present and future destructive growth patterns."[7] Others saw corruption in the firing of both Hicks and Stewart. It was "just another bit of evidence that shows the government . . . is a 'puppet on a string,' being controlled by land developers," one St. Petersburg woman wrote.[8] "We've had it," wrote another. "Highrises shutting out the view . . . and along the beaches, continued issuing of permits for big complexes that will drain us of our resources while fattening the purses of architects and developers— and now this last highhanded bit of business—firing the very people hired to guide us. . . . This time the electorate has had more than enough and we may yet have the last word."[9]

If they didn't exactly get the last word, citizens at least made their voices heard. St. Petersburg's city manager system made it hard to challenge Hicks's firing, and Harbaugh made it clear that he had no interest in rehiring Hicks. The city council had little power and less interest in forcing the matter despite citizens' protests. Hicks eventually started legal action under the provisions of the 1970 Clean Water Act that protected water quality managers from summary firing. The city rehired him in a different capacity in 1975.

Hillsborough County commissioners were far more vulnerable to the will of the voters. Most citizens agreed that Stewart was acerbic, but they liked that he stood up to polluters. His no-nonsense bluntness fit their mood. Dozens descended on the county's Environmental Protection Commission meeting a couple of days after the firing. Hundreds more piled into the chambers at the next full commission meeting. And they kept coming when the commission wouldn't concede to their demand that Stewart be rehired. *60 Minutes* ran a follow-up segment, "Go Fight City Hall," to let viewers know about the fallout from their earlier reporting. The county commissioners who opposed Stewart dug in their heels and refused to rehire him. Stewart stayed in the news all summer. He gave speeches to civic groups, answered questions from reporters, and mulled the idea of running for a seat on the commission. Like Hicks, he also contacted the U.S. Department of Labor and threatened to sue for his job.[10] The County Commission voted three to one to rehire Stewart on 13 August when the attorney hired to represent the board told them their case was indefensible. It was just as likely councilmembers feared the voters. In the words of *Tampa Tribune* fishing columnist Herb Allen, "Seems that everyone hated Roger Stewart except the people."[11]

A diverse group of people stood up for Roger Stewart and Emil Hicks in 1974. Students, housewives, young professionals, fishermen, and scientists showed up at meetings and wrote letters of support. Garden clubs, civic organizations, and conservation groups offered them venues in which to tell their stories. Even some politicians spoke up in support of the two men. By 1974, the environment had become a critical issue of concern for Floridians, just as it had for Americans generally. Air pollution, water pollution, noise pollution, light pollution—all products of the nation's growing postwar economy—signaled to many Americans that the high standard of living they enjoyed had its costs, and that those costs, in turn, reduced the very comfort, health, and beauty that prosperity was supposed to deliver.[12] Informed by the budding science of ecology, inspired by the success of protest movements like those for civil rights, and convinced that government had the right, power, and obligation to improve people's lives, Americans joined a wide range of organizations to pressure political leaders to fix the problems. In response, national, state, and local leaders passed a raft of laws in the 1960s and 1970s that protected air and water by establishing standards for their health and providing resources to clean them up.

Tampa Bay became the centerpiece of local environmental organizing in this era. Its problems were obvious and widespread. People could smell the air along Bayshore Boulevard. They could see the sewage at Spa Beach. They knew they couldn't eat the oysters, that they caught fewer fish. They saw developers build houses and condominiums and port facilities on top of seagrass meadows and mangrove forests. Health, leisure, aesthetics, and economics met at the water, and the view from the bay revealed the connectedness of it all. The problems in the water, on the land, and in the air were one. Residents could see the tragedy unfolding around them, so they organized to do something about it. But organizing was just the first step. There were battles to be fought, too. Land developers, shipping interests, and phosphate companies dominated local politics and resisted anything they felt would cost them money or constrain their activities. Political leaders, committed to low taxes, balked at proposals to spend on infrastructure, particularly the kind people need but rarely see. Smooth roads and new schools impress voters in ways that new sewage plants do not. And there were intersquad skirmishes, too. Fishermen, homeowners, biologists, and environmentalists could all agree that the bay needed to be cleaned up, but their interests sometimes clashed, especially when it meant that one group might have to bear a greater burden on their lifestyles in the name of cleaning it up. Nevertheless, in Tampa Bay they found a cause they could rally around.

It wasn't as if people hadn't noticed problems around the bay before the postwar boom years. More than a half century earlier, conservationists advocated for laws to protect its wildlife. The Audubon Society, formed as the American Ornithological Union in 1886, pressed for the passage of state and federal laws to ban the feather trade. Conservationists, whose numbers included many upper-class clubwomen, founded the Florida chapter of the Audubon Society in Maitland in 1900. The following year, the Florida legislature outlawed the killing of nongame birds and made it a crime to possess or sell any part of them or molest their nests or eggs. With few wardens—the state didn't fund the hiring of any, so Audubon employed as many it could afford—violators went uncaught. Six months after passage of the law, observers noted that the "wholesale destructions of birds at Bird Key" in Boca Ciega Bay continued unabated.[13]

Blatant violations of the bird laws nationwide convinced President Theodore Roosevelt to act where he could. He declared Pelican Island, in the Indian River Lagoon on Florida's east coast, the first federal bird sanctuary in 1903. Two years later, after much pleading from Florida Audubon leaders, he extended the same protection to Passage Key at the mouth of Tampa Bay. The Audubon Society appointed boatbuilder Asa Pillsbury as warden, a job he and his wife apparently both pursued with vigor. A couple of years after their appointment, an estimated 1,500 tricolored herons nested in the mangrove forest at the island's north end, protected by the Pillsburys' watchful eyes and ready rifles.[14] In 1906, Roosevelt added Bird Key (identified as Indian Bird Key at the time) to the growing list of bird sanctuaries. In 1921, thanks to the efforts of Katherine Tippetts and the St. Petersburg Audubon Society, President Warren G. Harding added Tarpon Key and some other islands to the Bird Key preserve when local conservationists pleaded to save the islands from private development.[15] In addition to those protected areas, the Florida Audubon Society hired a warden to protect the birds of Tampa Bay's eastern shore, particularly on Whiskey Stump Key and Green Key, just south of the Alafia River in an area known as the Kitchen. But as late as the 1930s, they continued to find "every evidence of vandalism and wanton destruction of life" carried out by "poachers of various kinds . . . egg collectors and gunners shooting into the rookeries for sport."[16]

While the Audubon set pursued conservation for aesthetic and recreational reasons, commercial fishermen championed the first protections for the bay's fish in the name of protecting their livelihoods. As early as the 1890s, Cortez mullet fishermen first complained of reduced hauls. No one scientifically counted the fish in those days, so there is no record against which to measure, but it is not far-fetched to imagine that years of hauls numbering in the tens of thousands took a toll on fish populations. In 1897, state senator W. H. Fuller of Bradenton shepherded through the state's first fish law. It banned stop netting across bays and rivers, established maximum net lengths and minimum mesh sizes, prohibited net fishing between June and August, and created a closed mullet season that barred taking the fish between 15 November and 31 December.[17] It was, in the words of one Cortez fisherman, "a most excellent law."[18] Operators of Tampa's fish houses at first opposed it, fearing it would leave them with nothing to ship during the summer and early winter, but they changed their minds when it became clear that those were their least profitable seasons.[19] It may be, too, that they simply figured that, with weak enforcement, they could keep fishing. "No attention whatever is being paid to the fish laws of the state," a journalist noted in 1905. "Mullet are being caught and sold in the streets of Tampa and the famous 'stop net' . . . is reported as being used along the coast."[20] Struggles over enforcing net laws continued for decades.

Bird and fish laws, to which the legislature added laws protecting oysters and sponges from overharvest (but not pollution), represented a conservationist impulse toward nature. Most conservationists took a practical approach: what value, even aesthetic value, do these things have for people and how do we make the wisest use of them? A fight over pelicans that pitted bird lovers against recreational fishermen revealed how answering such questions could engender conflict even among those committed to conservationist ideals. Fishermen around Tampa Bay hated pelicans, seeing them as competition. In 1926, the St. Petersburg chapter of the Izaak Walton League called for the extermination of the area's pelicans. Founded in Chicago in 1922, the league championed conservation efforts that would ensure protection of game fish for recreational angling. Local fishermen started a Tampa chapter in 1924; St. Petersburg and Oldsmar chapters followed in 1925. League members called pelicans "ugly old murderers" that ate more than seven pounds of the fish a day that were better reserved for human anglers.[21] The St. Petersburg and state Audubon societies enlisted scientists to explain that pelicans ate mostly nongame fish, but anglers remained adamant. Fortunately for the pelicans, St. Petersburg's political leaders had no interest in allowing anyone to slaughter the city's mascot.

The argument over pelicans also revealed another limitation of conservationist thinking: the tendency to view the natural world in parts, usually with a focus on larger, charismatic fauna. It was only after the 1920s that the science of ecology, which views nature in terms of systems and emphasizes interconnectedness, came into its own.[22] Scientists increasingly took a systemic view of nature, attempting to understand how the parts worked together and how changes to part of the system could reverberate throughout. The development of ecological science coincided with the expansion of marine science, which received a boost during World War II because of increased need for knowledge of maritime worlds, both on and under the water. Together, these sciences brought new thinking about estuarine environments like Tampa Bay.

Writers and filmmakers, in turn, popularized the insights of ecologists and marine scientists. Rachel Carson's bestsellers *The Sea Around Us* (1951) and *The Edge of the Sea* (1955) persuaded readers of the interconnectedness of not only the seas, but of land and sea. (Carson's fame and influence on environmentalism increased further with the publication of her landmark book *Silent Spring* in 1962.) Similarly, Jacques Cousteau's film *World Without Sun* (1964) and his subsequent television series brought deep-sea science to life for viewers.[23]

Thanks in part to Cousteau and Carson, more people wanted to be near, on, or in salt water. Millions around the country took up sportfishing, boating, swimming, and diving. Tampa Bay seemed to be the ideal place to pursue life on the water. In 1966, there were 25,000 personal boats registered in the Tampa Bay Area, as many as in the entire Chesapeake Bay or Long Island Sound.[24] Demand for beaches and boat ramps and the fingers of suburbs reaching into the bay all reflected newcomers' desires for the promise of life by the ocean.

Pollution broke that promise. Between the smell on the Bayshore, the lack of fish in Hillsborough Bay, and the increasingly common beach closures due to sewage pollution, it was hard to see Tampa Bay as a jewel for watersports. Others complained of the "vulgar, sign-littered atmosphere . . . on the beaches" and the loss of natural areas to "hotdog stands."[25] "Have a beautiful view of the bay now," a Pinellas County woman wrote home to Indiana in 1959, "but, of course, they will build houses soon and all we'll see is back doors."[26]

Appalled at what they saw going on around them, beginning in the 1950s citizens organized to confront the consequences of growth. The dredging of Boca Ciega Bay was their first major fight. In 1954, Mary Bigelow, Floyd Brown, and Mrs. Robert Davis created the Pinellas Alliance for Conservation of Natural Resources.[27] Members of the new association were well connected locally. Bigelow was married to the owner of a Madison Avenue advertising firm that relocated to Pinellas County in the early 1950s. Davis was president of the Garden Club of St. Petersburg. Brown was a high school science teacher and president of the Gulfport City Council. Dismay over the prospect of the dredging of Boca Ciega Bay brought them together. In 1955, they reached out to the local Rod and Gun Club and various neighborhood associations for support.[28] The Alliance also took on formal affiliation with the Florida Wildlife Federation. The Alliance petitioned political leaders, organized their neighbors, and spoke out at any meeting held to discuss permitting for dredging projects. Mary Bigelow was known to "promote her cause with considerable fervor," one hand "cutting through the air" while the other clenched "a fat file folder full of maps, diagrams, and notes."[29]

The scientific facts, she and her allies believed, proved the foolishness of ripping up the bay bottom. Convinced the science would persuade the community, she hosted at least one gathering at her home on Boca Ciega Bay to have maps of the proposed fills explained by marine biologist Robert Hutton.[30] In 1956, Hutton and his team produced *The Ecology of Boca Ciega Bay, with Special Reference to Dredging and Filling Operations*, the first ecological study of any part of Tampa Bay, when the Florida State Board of Conservation requested a study of the effects of dredge and fill. Scholar R. Bruce Stephenson has argued that the study was "an important test case for a nation wrestling with the issue of environmental protection" because it employed the "subversive science" of ecology, which relied on models of natural communities evolving to efficient points of equilibrium. This viewpoint challenged assumptions that nature could easily withstand human-induced change.[31] Hutton's study considered the health of the bay's ecosystem holistically, offered projections of how alterations would change it, and argued that dredging would destroy not just parts, but the whole. Advocates of dredging accused Hutton and his team of using science as a cover for their own preference not to see the bay dredged, but they insisted that it was not.[32] The Hutton Report failed to convince state and local authorities to stop the dredging and filling of Boca Ciega Bay. By 1962, it was clear he had been right. Another scientist surveying the new development decried its "sterile sea walls" that left only "sluggish pools" where only "trash fish will live" because the water was "less and less enriched by washing nutrient-bearing shores and marshes."[33]

Activists had not yet convinced politicians (or most residents) of the need to protect the bay from development. But their voices grew louder. In 1965, St. Petersburg residents formed the Committee to Save Our Bays to protest additional dredging to build Tierra Verde near the mouth of Tampa Bay. The head of the project dismissed the group as people concerned "over the love life of a fiddler crab and a couple of clams."[34] He underestimated the growing popularity of the anti-dredge message. The committee's chairman, print shop owner Blake Davis, distributed more than 40,000 flyers opposing the expansion. On the back of each flyer was printed information for contacting county commissioners. Canvassers collected more than 2,000 signatures of those opposed to the plan in one weekend; only six refused to sign.[35] Within weeks, upward of 20,000 people signed and Davis was still printing flyers. One part-time resident offered that many of the New York snowbirds he knew wanted flyers to register their opposition. They had read about it in a *Sports Illustrated* story that condemned the expanded dredging.[36] They ultimately failed to stop the project, but opponents of dredging were getting through to the people, even if they weren't yet winning the fights.

A harebrained scheme to turn one of Tampa Bay's largest estuaries into a freshwater lake gave the fledgling environmentalists their first victory. In the postwar years, Florida politicians seemed to believe that, with help from the U.S. Army Corps of Engineers and enough earthmoving equipment, almost any environment could be changed to meet the demands of a growing population. The drained Everglades, the straightened Kissimmee River, and the work being done cutting the Cross-Florida Barge Canal across the state supported their faith in bulldozers. In the Tampa Bay Area, the most pressing need was for fresh water. While the area received more than four feet of rainfall annually and the limestone under the soil held even more water, localized water shortages were a fact of life. By mid-century, the water needs of citrus growers, cattle ranchers, and a growing suburban population had begun to tax that aquifer. In the 1950s, residents in the community of Odessa expressed fears that overpumping was creating sink holes in the limestone. The Bay Area's "water wars" raged for decades.

In late 1956, state engineers suggested that Upper Tampa Bay might make a good freshwater reservoir. Armed with aerial photos, they noted that its location made it ideal to serve both western Hillsborough and northern Pinellas Counties. Better still, they said, it could recharge the aquifer the counties fought over. Hydrographic charts, they said, told the story of an Upper Tampa Bay getting shallower and sicker because of the Courtney Campbell Causeway. The causeway was almost a dam as it was, they argued, why not just convert it and make Upper Tampa Bay more useful to people?[37] "Upper Bay Lake Project a Legislative Must," trumpeted the editor of the *Tampa Times* shortly after the Tampa Chamber of Commerce started the drumbeat for the project.[38]

The visionaries claimed conversion of bay into a lake would be easy: simply cut off Upper Tampa Bay and the natural inflow of fresh water would be enough, in time, to make it all fresh. They pointed to Pinellas County's Lake Seminole, which was made in the 1940s when the construction of Park Boulevard had cut off the brackish northern reach of Long Bayou. They failed to mention that Lake Seminole measured only 684 acres; the Upper Tampa Bay project would convert seventeen square miles.[39] As engineers examined things more closely, the simple became more complicated: Would there be enough fresh water entering the lake? Yes, but barely, the Army Corps of Engineers explained. How long would it take? No one could say for sure. Maybe five years? Maybe seven? Could the causeway even be turned into a dam? No, the open spans would need to be replaced with locks to maintain navigation between the bay and lake. Besides, the Courtney Campbell was not tall enough. It would need to be raised two to seven feet (no one was sure how much) to prevent salt water from going over the dam in storms. How much would it cost? No one could say.[40]

Despite the open questions, the Greater Tampa Chamber of Commerce, the Clearwater Chamber of Commerce, and numerous politicians pressed forward with planning in the name of fresh water. And as time went along their promises for the lake's potential only grew rosier: it wouldn't just solve the water problems, they said, it would bring more tourists, encourage more development, protect the houses along Upper Tampa Bay in hurricanes, and control flooding during seasonal inundations. Boaters could even enjoy it as a place to kill the barnacles that fouled their boats! Supporters met most objections with contempt. Those that couldn't be dismissed they turned over to the engineers who could, on paper, solve any problem. In 1961, the Southwest Florida Water Management District (SWFWMD, pronounced "Swiftmud") took the lead in overseeing the project's development as part of a larger flood control project in the Upper Tampa Bay watershed. The agency accelerated the planning process when it seemed the new lake could help with seasonal flooding around Pinellas County's Lake Tarpon. Politicians and local media cheered the agency on, confident engineers would overcome any impediment. In February 1965, SWFWMD approved moving forward with the project.[41]

While engineers told politicians the project *could* be done, a growing chorus of constituents asked whether it *should* be done. There had been objections from the beginning. As soon as the plan was announced, residents along Upper Tampa Bay voiced opposition, fearing it would ruin everything they loved. "We want the bay kept just like it is," Safety Harbor's mayor explained in 1958, adding that 95 percent of the town agreed with him.[42] SWFWMD's 1965 approval signaled opponents like a fire bell. Oystermen, who had already been dealing with the problems of siltation and pollution and for whom the conversion would mean the certain death of Tampa Bay's last viable oyster reefs, protested the economic effects of the conversion. Upper Tampa Bay's crabbers agreed with them. It "would hurt like hell," crabmeat processer F. H. Hartman told a reporter.[43] Everyday citizens added their voices. In April 1965, the Alliance for Conservation of Natural Resources and the Committee to Save Our Bays—the twin defenders of Boca Ciega Bay—joined forces to stop the damming of Upper Tampa Bay and created the Save Our Bays Committee.[44] Members bombarded state and federal officials with "a fusillade of wires" and kept up the pressure in the press and at meetings of county and SWFWMD officials.[45] The Florida Division of the Izaak Walton League registered its opposition in petitions to the Army Corps of Engineers, Governor Haydon Burns, and Florida's congressional delegation.[46] An aide to Representative Sam Gibbons noted that the Army Corps of Engineers office received "a tremendous number of protests" regarding the lake.[47]

Academic scientists played a key role in supporting opposition to the so-called Bay Lake plan. Scholar-activists, they refused merely to study the ecological problems they saw around them and instead took up the work of teaching the public about what was happening. Their credentials legitimized their concerns while their expertise, both as scientists and teachers, helped residents to think in new ways about the bay's ecology. Most taught at the new University of South Florida (USF) in Tampa (founded 1956), the newer Florida Presbyterian (now Eckerd) College in St. Petersburg (1959), or at the area's burgeoning community colleges. Most were young (or at least youngish) and on the leading edges of their fields. Dr. George K. Reid, professor at Florida Presbyterian, was a pioneer in the field of marine ecology who wrote one of the first textbooks on estuary science, *Ecology of Inland Waters and Estuaries*.[48] He and his colleagues offered the college's first course in marine ecology, which they opened both to students and "anyone . . . otherwise interested in the subject."[49] Dr. William H. Taft came to USF in 1963 after writing a dissertation on Florida's coastal geology at Stanford; he spearheaded the creation of the university's oceanography program. Dr. Sylvia Taylor was an expert in algae who taught marine biology at St. Petersburg Junior College and went on, as Dr. Sylvia Earle, to become one of the world's most revered oceanographers.[50]

Those scientists and their colleagues excoriated the Upper Tampa Bay project. Taft, the most vocal of the scientists, pointed to errors in the bay bottom samples and concluded the idea "would never have matured past the talking stage" if a geologist had reviewed it.[51] Reid agreed and predicted that it would be "one big, muddy, stinking mess" without a source of water that would outstrip evaporation. Taylor added that the shallow water in the cutoff section of bay would get as warm as ninety degrees Fahrenheit during the summer, supporting algae blooms that would cause regular, odorous fish kills. And that was in addition to the widespread "offensive odors from rotting plants and animals" killed by the conversion. Not only would it stink, an entomologist at the State Board of Health added, the area's mosquito problem would get worse as disease-carrying freshwater species replaced their merely annoying saltwater cousins.[52] Fisheries scientists, meanwhile, pointed to the bigger picture. They assessed that the project would kill "one of the most productive" parts of Tampa Bay for "young commercial fishes."[53] Upper Tampa Bay was a link in the "chain of life" in the Gulf of Mexico, the regional director of the U.S. Bureau of Sport Fisheries and Wildlife explained, and "if you choke it off in one part of the chain, you choke it off in the Gulf."[54]

Gibbons, Hillsborough County commissioners, and much of the Tampa business community doggedly held onto the lake idea. Hillsborough County Commissioner Frank Neff described the anti-lake pamphlets that Blake Davis distributed at one commission meeting as "scandal literature."[55] The *Tampa Times* disparaged "Dr. Taft and his small band of dissidents."[56] Local CBS affiliate WTVT Channel 13, the Bay Area's oldest television station, dismissed opponents as malcontents who couldn't present "a legitimate and convincing objection" to the plan.[57] Alfred McKethan, chairman of SWFWMD's board, dismissed objections as "mostly bunk."[58]

But the scientists' campaign to educate the public gave energy to the environmental movement. By the fall of 1965, Save Our Bays had convinced Pinellas County commissioners to back out of their support for the lake.[59] In 1966, the newly chartered Tampa Audubon Society sponsored a series of public information sessions that aired arguments both for and against the project but tended to build opposition. By August 1966, State Conservation Director Randolph Hodges signaled his opposition to the plan. "I don't think we can sacrifice any more of our salt water resources," he told reporters.[60] A month later, Hillsborough commissioners, sensing flagging support, removed funding for the project from the county's budget. The most ardent supporters held out hope that the project might be revived, but in March 1968 officials at SWFWMD moved to abandon it and focus on flood control.[61] Later that year, Reynolds, Smith and Hills, the engineering firm hired by SWFWMD, fulfilled their contract and turned out a formal proposal for the bay lake.[62] It was a dead letter.

The battle over Upper Tampa Bay wasn't over, though. In the late summer of 1969, a developer pitched an idea to build a massive community on its last large tract of shoreline: the Bower Tract. Named for owners who purchased it in the 1950s and located east of Oldsmar, the tract's 3,000 acres stretched from Rocky Creek in the east to Double Branch Creek in the west. Only about half to two-thirds was dry land, though, depending on where one considered the bayside property line to be. (It was a matter of dispute.) The shoreline was a mix of mangrove forests, salt marshes, and seagrass flats that gently sloped into the water. Small tidal creeks divided it in places. At high tide, water covered most of the shoreline; at low tide, it was muddy but walkable. Developers had first pitched the area's potential before the land boom crashed in the 1920s. In the 1950s, state officials eyed it as a potential home for USF but balked at the cost of developing the property. In 1959, the Tampa Shores Company announced it would dredge and fill the property and build 12,000 houses on dozens of fingers jutting into the bay.[63] The plans proved too ambitious; crews dredged a few fingers on the east side of Rocky Creek and built several dozen houses, but construction soon lagged. As late as the 1970s, Tampa Shores had closer to 120 houses, not 12,000.

Despite the failure of Tampa Shores to take off, development company Intervest announced plans to build a project on the shores of Tampa Bay "unlike anything . . . on Florida's West Coast." Intervest pitched its Bay Port Colony as a "totally planned community" that would include a yacht club, a golf course, office buildings, retail space, a convention center, and housing for eighty thousand people! Artistic renderings depicted multiple condominium buildings rising higher than ten stories over the bay. All housing units, they promised, would have a water view over either the bay or the twenty-three miles of canals they would dredge. They planned to price houses between $35,000 and $40,000—expensive, but not out of reach for well-to-do retirees or white-collar professionals. They promised to spend $1 billion ($8.5 billion today!) and more than a decade on the project.[64] The plan was bold but not out of line with the dozens of other development schemes being pitched around Florida in that era.[65]

Given the controversies over dredge-and-fill projects and waterfront development generally, Intervest's owners no doubt expected nature lovers to chirp at their project, but they were counting on elected officials to back their plans. Boca Ciega Bay was a testament to developers' ability to charm politicians. So was Apollo Beach, the small city on finger islands cut out of the mangroves on the bay's eastern shore beginning in the late 1950s. There, developer Francis Corr faced a thousand problems in getting the planned community off the ground, from rutted roads to recalcitrant rattlesnakes, but he couldn't blame environmental regulations for the remote community's slow growth.[66] Closer to the Bower Tract, Dana Shores and Bay Crest Park, both north of the east end of the Courtney Campbell Causeway, also proved the ease with which the shoreline could be transformed in the postwar years. When developer Alfred Dana finished dredging the finger islands of his eponymous development in 1957, the *Tampa Times* praised him for "enhancing the countryside" and "providing Tampa with something it sorely needed, more waterfront property—the kind that could be used as homesites."[67] The visionaries at Intervest saw themselves upholding that tradition.

What they didn't understand was that the changing mood around the Bay Area had changed the political environment. It wasn't that the Bower Tract was a pristine waterfront. It was already showing the effects of runoff pollution, and flood control engineers had just cut a drainage canal through the heart of it. Intervest, in fact, promised to build a state-of-the-art sewage system that would take up some of the load of the inadequate systems plaguing nearby developments. Instead, the Bower Tract was the right place, and Bay Port Colony came at the right time, to crystalize the frustration with the state of Tampa Bay into a movement to save it.

As with the Bay Lake, scientists led the charge. The dredging and filling concerned them the most. Intervest planned to fill in more than 1,700 acres of low-lying land to a height of 7.5 feet above sea level using 10.7 million cubic yards of fill, much of it drawn from adjacent bay bottom to which they laid claim.[68] The state's recently passed Register Act (named for Pinellas County state representative William Register, who also happened to work as an attorney for Intervest) required a biological survey of the impact of any dredging before it could proceed. The Florida Department of Natural Resources' report was damning. Dredging the area would have "massive adverse effects on marine biological resources in Tampa Bay." The project would disturb an area "rich in wildlife, marine life, and birds" and eliminate "productive nursery and . . . [feeding] grounds for marine life and wildlife."[69] Roger Stewart, at that time Hillsborough County's chief biologist, was blunter: allowing the dredging would be a "gross mistake" that could "spell the end of Old Tampa Bay."[70] University of Florida marine biologist William Carr, who, when he learned of the plan, descended on the Bower Tract in early 1970 with a small team of students to collect fish specimens and take core samples, declared that allowing the fill "would be an environmental abuse I cannot comprehend."[71] It seemed that Dr. William Taft popped up at every meeting about the plan to point out how bad the project would be for the health of the bay. He "lurked in the mangroves like a Viet Cong terrorist, armed not with Molotov cocktails, but with bottles of polluted water," one reporter remembered.[72]

Their warnings galvanized opposition. Letters poured into the local papers while citizens piled into meetings about the project. Opponents to the dredging came from many circles. Sport fishermen were especially vocal. The Golden Triangle Sports Fishing Club called the project a "crime against nature." Its survey of more than 1,200 attendees at the Tampa Boat, Travel, and Sportsman's Show found one person in support of it.[73] The president of Interbay Marina took out a large ad in the *Tampa Tribune* encouraging citizens to pressure the Port Authority, which controlled the permitting process, to stop "this proposed rape of our natural resources."[74] Commercial fishermen joined the chorus opposing the development, fearing it would "destroy the prime nursery areas for fish and shrimp on the West Coast."[75] Local garden clubs, the Audubon Society, and hundreds of college students from USF, Hillsborough Junior College (now Hillsborough Community College), Florida Presbyterian, and other schools also held demonstrations against the development. At USF, the student group Environment 70 organized a telephone call campaign as part of the university's inaugural Earth Day program.[76] "They don't care about us," one seventeen-year-old from Dana Shores explained of her opposition to development. "They're ruining the environment for our generation."[77]

In no small irony, much of the opposition to the plan came from people who, like this student, lived in other bayfront developments. It would be easy to chalk up their opposition to NIMBYism or write it off as hypocrisy, but their concerns were rooted in their own experiences. They understood the consequences of unchecked growth because they could see it out of their picture windows. "The first year we were here, the water looked fine. We swam in it all the time," Dana Shores resident John Kitko explained in 1970. "The second year it began to look dirty, and last year, it looked and smelled so bad it was apparent that something was wrong."[78] Angered by their inability to get county leaders to respond to the effluent from inadequate sewage treatment plants in their back yards, Kitko and members of the Dana Shores Civic Club reached out to civic associations in neighboring subdivisions and held informal meetings with neighbors from Bay Crest Park, Rocky Creek, and Tampa Shores. In 1969, they formed a new Save Our Bay Committee. While it had a similar name to the group that had challenged the Bay Lake, it was a different organization, and it is not clear if its organizers knew of the earlier group. One member later recalled that they chose the name because it fit their mood. Developers "call us s.o.b's, so how about Save Our Bay?"[79] When news of the Bay Port Colony broke, they broadened their focus. They worried that it would be a massive new source of pollution that would doom their efforts to clean things up. "When the Bower tract came along, it united everyone," a Save Our Bay member explained.[80]

Most members of Save Our Bay were young professionals: college professors, engineers, attorneys, stockbrokers, at least one minister, and several military officers were among its leaders. Men garnered most media coverage, but women played a significant role in its activities, including holding seats on the organization's board of directors. Betty Castor, who later went on to serve on the Hillsborough County Commission, was president of the Hillsborough County League of Women Voters; she joined the group even though her Carrollwood home was nowhere near the bay. Her husband, Donald, an attorney with considerable political connections, served as the organization's first treasurer. Both had played an important role in publicizing the 1969 Hagen Report about Hillsborough Bay by organizing a seminar entitled "Who Killed Tampa Bay?" ahead of the report's release.[81] John Kitko, a naval commander assigned to MacDill Air

Force Base after a stint in Vietnam, served as the group's first spokesman, but he was quick to tell people that Save Our Bay initially had "no formal organization and no officers." "We sit and drink coffee and talk it over and then, when we decide what needs to be done, whoever can do it offers to help."[82] Save Our Bay was no mere coffee klatch, though. In February 1970, the group convened a public meeting with the county's pollution control committee at which more than three hundred angry residents met county commissioners with "boos, catcalls, pressure and pleas" to clean up the bay. They came from all over the Bay Area to tell their stories of frustration with the problems of pollution. It went on for five hours.[83] In the summer of 1970, Save Our Bay incorporated as a nonprofit organization to better manage fundraising for its activities. Its leaders claimed that more than 7,000 people joined.[84]

Activists brought unprecedented scrutiny to the Bay Port Colony project. In legal filings, Donald Castor revealed that nagging questions about land titles and boundaries hung over the entire project. It wasn't clear whether Intervest could hold legal title to submerged lands, whether the port authority still had the authorization to allow dredging projects, or whether Hillsborough County, where most commissioners were sympathetic to the developers, or the City of Tampa, where most members of the city council opposed the project, controlled the permitting process.[85] And that was before the Army Corps of Engineers even began its assessment of what the dredging meant for navigation. All those problems might have been overcome by builders determined enough to keep pushing, but Intervest announced it was dropping its request for permits. "The ecologists went berserk," Intervest president Bruce Clary later explained to reporters, and the company was not willing to risk its shot at a permit in that political environment.[86]

Clary and Intervest's leaders assumed the winds would shift again, the protests would calm down, and things would be back to normal. They were wrong. In January 1971, the company announced it was still planning to develop the Bower Tract. Two months later, the Tampa Port Authority announced that dredge and fill operations could not extend past the mean high-water line, severely constraining Intervest's plans. That summer, when Hillsborough County commissioners approved work on the upland portions of the land, the company fired up the dredges to dig unapproved canals. They apparently hoped to push the project far enough along and force the various agencies to allow them to continue. And in an earlier era, this strategy might have worked. But Save Our Bay and other activists were watching. Save Our Bay sued and sought an injunction against the project.[87] Lawyers for the federal government followed on quickly, charging violations of the Rivers and Harbors Act.[88] In the end, Intervest had to return all the dirt they had dug out. It was a first. Shortly after, the company announced it had purchased land inland from the Bower Tract for future development and would not be developing the land near the water.[89] In 1985, after much wrangling, the state purchased the Bower Tract to protect it from future development.[90]

Dredging, pollution, the Bay Lake, the Bay Port Colony, and dozens of other major and minor assaults on the bay led Bay Area residents to find a cause in the health of Tampa Bay. For some, the idealism of the environmental movement proved enough reason to care, but for most, it was the material problems they confronted—sulfurous air, fouled swimming areas, unhealthy canals—that led them to demand more from their elected leaders. It would be a hard fight.

7

ESTUARY

Just after sunrise on 13 February 1970, the oil tanker *Delian Apollon* struck a submerged spoil pile as it neared Florida Power's fuel port on Weedon Island. As tugboat operators struggled for more than an hour to free the ship, 20,000 gallons of fuel oil spread across ten square miles of Tampa Bay south of the Gandy Bridge.[1] Port authorities, lacking equipment, spread straw to control the spill. The heavy oil stymied their efforts, just as it defied the divers who tried to plug the holes in the ship with rags and blankets. Winds and currents spread the oil south and west. A day after the collision, it reached Point Pinellas. A day later, it washed ashore in Boca Ciega Bay and oozed as far north as Pass-a-Grille. Volunteers sprang into action, washing ducks, spreading straw, and doing anything they could to help with the cleanup. For weeks, people found birds covered in oil as far south as Manatee County. Ultimately, cleaning up the spill cost an estimated $10 million (roughly $80 million today), making it the most damaging oil spill in Florida's history to that point.[2]

The spill made national news. A year earlier, a catastrophic spill in Santa Barbara, California, had shocked Americans into the realization that even the nation's beautiful places were not safe from pollution. The *Delian Apollon* reinforced the lesson. In March, *Life* magazine ran an eight-page article with damning photographs. An oil-suffocated duck floating on a sheen of black liquid covered two pages. Other images included oil-splattered mangroves, beaches covered in oil-soaked straw, and volunteers chasing down affected birds to wash them.[3] Understanding the bad publicity, the state's lieutenant governor and attorney general both sued Florida Power, Humble Oil, and the owners of the *Delian Apollon*.[4]

Business and political leaders around the Bay Area typically would have lamented the spill but chalked it up as the cost of doing business. Exactly a week before the *Delian Apollon* ran aground, workers at the Borden phosphate processing plant on Piney Point spilled acidic water into Bishop Harbor, killing thousands of fish. The company never reported it. Two weeks later the company spilled more acidic water—landing what Manatee County's environmental engineer called the "coup de grace of Bishop's harbor" [*sic*]. The state's pollution control agency could do nothing more than investigate and document the event since it did not directly affect public health. Dead fish were the cost of prosperity.[5]

The *Delian Apollon* spill dramatized the assault on Tampa Bay in ways that your average fish kill could not. "Suddenly," the editor of the *Tampa Tribune* noted, "everyone became pollution conscious."[6] He exaggerated, of course, but he also unwittingly saw the future. Beginning in the early 1970s, the ethos of life around the Bay Area began to shift, albeit slowly. Tampa Bay's environmental problems grew too obvious to deny and citizens demanded change. Scientists, environmentalists, and everyday folks worked hard to convince political leaders and the business community that the bay had been pushed to its limits and the demands on it needed to be rebalanced. In 1972, Jan Kaminis Platt, who later as a Hillsborough County commissioner would become the bay's greatest champion, spoke for many in a letter to the *Tampa Tribune*. "The time has come when we who inhabit this beautiful universe are awakening to the realities of not only the limitations we as humans can place upon that beauty, but also the limitations we are placing upon our very existence. Ways must be devised whereby the government can fulfill its responsibilities and industry can function efficiently within a framework compatible with preserving our natural environment. If a conflict exists among the three, then environmental considerations should come first, lest we as humans be last."[7]

Platt's framing of Tampa Bay's health as a problem of stretched limits and its salvation as a question of balancing interests mirrored the science of estuaries in those days. As scientists studied Tampa Bay in this era, they highlighted the key insights of estuary science: connectedness and balance. Taught to understand the bay around them in these terms, environmentalists used connectedness and balance as a metaphor for how they might proceed. Cleaning up an oil spill or stopping outlandish projects like the Bay Lake could rally people to action, but the questions of how to keep the ship channels dredged without incurring more ecological damage or how to press cities to improve their sewage systems or how to stop businesses from polluting required long-term commitments and negotiation. Helping people understand Tampa Bay as an estuary made it easier to show people where their interests might overlap and how they might work together when their needs conflicted. The metaphor nurtured a language of cooperation, even if it did not solve all conflicts. Cleaning up Tampa Bay took decades and a lot of hard work. But once business leaders, local governments, and everyday people could see that they were just as connected to the Tampa Bay estuary as the fish and the birds and the seagrasses and the mangroves, it created a framework for restoring Tampa Bay that has proved durable.

In the early 1970s, the Florida legislature and the U.S. Congress passed a wave of laws in response to citizens' increasing environmental awareness. The 1967 Florida Air and Water Pollution Control Act established the first legal standards for defining the extent of pollution in the state's waters and created an agency to investigate its causes.[8] The 1969 National Environmental Policy Act (NEPA), which President Richard Nixon signed into law on 1 January 1970, mandated scientific assessments of the environmental impact of major building projects. Congress followed with the Clean Water Act (1972) and the Endangered Species Act (1973), which included the West Indian manatee, the bald eagle, and the American alligator, all native to Tampa Bay's watershed. In 1972, the state legislature passed three landmark bills: the Florida Water Resources Act, the Comprehensive Planning Act, and the Land Conservation Act.

In accordance with the new laws, local governments and private companies hired scientists to study nearly every aspect of the bay. The construction of the Tampa Electric Company's Big Bend generating plant in 1970, for example, required the company to hire a team of marine scientists to study the effects of hot water discharge on everything from nesting birds to bottom-dwelling invertebrates. (This was long before anyone knew how attractive the heated water would be to wintering manatees.) Building on Robert Hutton's pioneering studies of Boca Ciega Bay and the Hagen Report's findings about Hillsborough Bay, scientists created a comprehensive picture of the bay's health. No doubt some reports reached conclusions that reflected the interests of the companies that paid for them, but more often the scientists collaborated with the growing community of academic scientists from the USF and other academic institutions and government agencies studying Tampa Bay. In 1976, the City of Tampa, looking to monitor pollution levels in Hillsborough Bay, created the Bay Area Study Group to support the sharing of data and to foster communication among scientists. It continued its work until 2009.[9] The close, long-term relationships forged by scientists from a range of fields further deepened local knowledge and reinforced in their minds people's need to see the bay as a connected system linking water, plant life, animals, and people.

The scientists' work informed the Bay Area's environmentalists, who themselves were becoming better organized. Energized by the battle to stop the Bay Port Colony, they pushed local politicians to do more about pollution. When a county commissioner offhandedly dismissed the concerns of Save Our Bay members by explaining that he was "interested in people" not "estuaries [and] marine life," one Riverview resident responded that "the people are interested in crabs and fish; they're not interested in (more) pollution."[10] "The county is not doing enough," complained resident Dennis O'Madigan, who accused Tampa Shores and Baycrest of dumping raw sewage into Old Tampa Bay. "I don't see any of those (sewage treatment) plants trying to get in line."[11]

Responding to similar complaints from around the bay, two Pinellas County politicians decided to make sewage plant operators clean up their act. In 1972, State Senator Harold Wilson and State Representative Mary Grizzle steered through a bill to end the dumping of sewage into Tampa Bay and surrounding waters. The Wilson-Grizzle Act mandated that all sewage in Hillsborough, Pinellas, Manatee, Pasco, and Sarasota Counties had to undergo advanced wastewater treatment before it could be released into any surface waters. This so-called tertiary treatment would remove not only solid waste, but also stop bacterial pollution and reduce the astounding volume of nitrogen discharged daily into the area's coastal waters. Several local governments, the Tampa Bay Regional Planning Council, and business leaders complained about the cost of meeting the law's requirements and opposed its passage.[12] But Wilson and Grizzle knew this was just the sort of bold action many in the Bay Area hoped to see from their elected leaders.

Despite grumbling, communities stepped up to follow the law. As part of the 1972 Clean Water Act, the federal government provided funds for local governments to improve their wastewater treatment facilities. Communities all over the Bay Area took advantage of this funding to upgrade their systems or develop alternative disposal methods. St. Petersburg rerouted processed effluent into water recycling programs like irrigation and deep well injection sites. In 1978, the City of Tampa opened the Howard F. Curren Advanced Wastewater Treatment Plant on Hooker's Point, a state-of-the-art facility designed to turn out pure water. At the opening ceremony, political leaders toasted champagne flutes filled with the treated water. Only Roger Stewart drank it. The chlorine in the water burned his mouth.[13] Within a year, scientists could see measurable improvement in the amount of nitrogen in the upper portions of Hillsborough Bay.[14]

As bold as it was, the Wilson-Grizzle Act addressed only one of the items on the long list of threats to the bay. In 1973, a coalition of environmental and civic groups, including Save Our Bay, the Tampa Audubon Society, the Sierra Club, the League of Women Voters, the Davis Islands Pollution Organization, and the Junior League, created the Hillsborough County Environmental Coalition to concentrate their activities in addressing the region's problems.[15] It joined a list of similar groups working around the bay and across the state, including ManaSota-88 (founded 1968), Suncoast Active Volunteers for Ecology (SAVE, founded 1970), and Conservation 70s (founded 1970). They planned to use the new federal and state laws to challenge dredge-and-fill projects, sue for better pollution controls, demand more from elected officials, and even promote viable candidates for political office. There was a meeting of environmentally minded citizens going on somewhere around Tampa Bay nearly every night of the week in those days.

The media environment also began to shift. The *St. Petersburg Times* expanded its coverage of environmental issues and regularly used its editorial page to criticize untrammeled growth. The *Tampa Tribune* and *Tampa Times* came around more slowly. Outdoors columnist Joe Guidry was among the first to address issues confronting the bay; other outdoors writers, including Frank Sargent, followed. Editorial pages also shifted, but the business page remained something of a cheerleader for the Port of Tampa's interests. Nevertheless, no one reading a newspaper in those days could have remained unaware of the bay's problems and the organizing people were doing to save it.

The port posed probably the thorniest problem for environmentalists. The local economy depended on a busy port, but its ecological impact was obvious. Ships leaked a witches' brew of chemicals from their fuel tanks and their holds. George S. "Skip" Gandy recalled seeing dying "seagulls thrashing around in . . . [and] tarpon rolling through the oil" along the seawalls of Davis Islands. Phosphate dust poured into the water from the loading docks, and processing plants on Hooker's Point dumped all manner of chemicals into the water.[16] By the late 1960s, large-scale building projects exacerbated the problems. Like port operators all over the world, the Tampa Port Authority had to adapt to a revolution in modality and size. Container ships, first perfected in the 1950s, replaced traditional break-bulk cargo ships, requiring port designers to rework their facilities to accommodate more trains and trucks as well as larger ships. In bulk cargo shipping, the method by which Florida's phosphate left Tampa Bay for the world, the trend was toward larger ships that sat deep in the water. To meet these changes, officials at the Tampa Port Authority expanded the port by building up Hooker's Point and constructing the East Bay Terminals. In doing all this work, they nearly killed McKay Bay by decimating its shoreline communities, nearly closing off its tidal flow, and concentrating pollutants.

Tampa Port Authority leaders pushed the U.S. Army Corps of Engineers for deeper and wider channels to handle the increased traffic they hoped would support the port's expansion. In 1974, Congress appropriated the funds to deepen the bay's channels from thirty-four to forty-three feet while widening the main channel by one hundred feet to bring it to six hundred feet. The Army Corps of Engineers expected to produce 72.3 million cubic yards of spoil (enough to fill as many as 7 million dump trucks). In lower Tampa Bay, they would dump most of the spoil on submerged piles parallel to the channel, but in Hillsborough Bay, they would pile it on five islands built using limestone boulders blasted out during the dredging process.[17]

Everyone knew dredging would be messy. More bay bottom would be irreparably altered (bad news for seafloor organisms and the things that ate them), and the act of dredging would necessarily churn up silt and make the water turbid (bad news for fish and seagrasses). No one was quite sure what pollutants might be released when dredgers dug up the pudding-like muck lining the bottom of Hillsborough Bay or what effects dynamiting the limestone base might have. Still, business leaders, politicians, and the Army Corps of Engineers reassured the public that the benefits would far outweigh the costs. At a press conference to announce the project in April, a dredging company paid hard-hat-wearing construction workers to line the walls of the Tampa Port Authority headquarters holding identical handwritten signs reading, "Opportunity, Prosperity. No

More Delay. Deepen Tampa Harbor" and "Jobs Provide Good Quality of Life. Tampa Harbor Provides Jobs."[18]

The Bay Area's environmentalists objected. Concerned that the Corps "would probably conduct this project in the tradition that has earned it a notorious reputation for ravaging the earth," local radio station WDAE (not typically known for agitating editorials) encouraged opponents to show up at the meeting.[19] More than seventy did, despite the fact that it was held at 9:00 a.m. on a Tuesday. Surrounded by the construction workers, scientists excoriated the project's engineers for treating nature as an afterthought. Frank Dunstan, the National Audubon Society's warden and biologist for the bay, called the Corps' environmental impact study "incompetent and inadequate." USF biology professor Joseph Simon, a specialist in sea worms who had begun studying Tampa Bay when he arrived at the university in 1964 (friends later called him Mr. Tampa Bay to honor his lifetime of studying its biology), pointed out that the species inventory included one worm "found no farther south than Long Island Sound" and another "that doesn't grow north of Miami." Roy R. "Robin" Lewis III, a Hillsborough Community College biology professor, vice president of Save Our Bay, and president of the Hillsborough County Environmental Coalition, criticized the study's assumptions about the long-term effects of dredging on sea life. USF geology professor Robert Taft warned of the potential of releasing pollutants like pesticides and heavy metals that had settled into the muck at the bottom of the bay. He worried, too, that the study was little more than a rush job created to satisfy new federal environmental guidelines. "Something's going on," he said of the absence of representatives of state and federal agencies charged with protecting the environment at the meeting. Only staff members of the Hillsborough County Environmental Protection Commission, the agency Roger Stewart had headed until his firing just weeks before, showed up to demonstrate their objections to the plans.[20]

After the "barrage of complaints" at the press conference, Colonel Emmett Lee, the district engineer for the Army Corps of Engineers' Jacksonville district office, which oversaw all projects in Florida, made an "on-the-spot decision" to empanel an advisory board of local scientists and other environmentalists from the Bay Area. He made no promises about the panel's power to alter the Corps' plans but said he would consider their ideas.[21]

It is not clear why Lee decided to open the door for the environmental-ists, but if he did it to co-opt and silence them, he miscalculated. Save Our Bay and the National Audubon Society took the lead. Audubon, which maintained several islands at the mouth of the Alafia River as bird sanc-tuaries, had worked for nearly a decade with the Tampa Port Authority to ensure maintenance dredging of the existing channels benefitted the bay's nesting terns. Save Our Bay had joined them in those efforts in the early 70s when Dunston and Lewis had worked together in studying the ecology of the bay's existing spoil islands. Their shared concern was that the new spoil islands would be too low above the surrounding water to be-come viable rookeries (a benefit the Corps touted), would erode quickly, and were positioned in a way that would further obstruct Hillsborough Bay's already weak tidal flow. They also worried that no plan was in place to ensure that the spoil was free of pollution before it was deposited.[22]

The issue proved contentious. The Corps was sure of its plan; the envi-ronmentalists were equally sure it was wrong. Lewis brought aerial pho-tographs and tide charts, first to the Corps, then to the Chamber of Com-merce, to show the weaknesses of the design.[23] Discussions went on for more than six months. At one point, Lewis suggested that Save Our Bay and the other organizations might take legal action to stop the project.[24] Spooked by the possibility of further delay, Representative Sam Gibbons facilitated a face-to-face meeting between Colonel Lee, Lewis and Dun-ston.[25] The meeting didn't get everyone on the same page immediately, but it improved the dialogue. The Corps soon agreed to adjust the islands' de-sign with input from Lewis and Dunston; these islands have since become important linchpins for restoring the bay's ecology.

The harbor deepening project took five years to complete. The Corps' records of that era are filled with correspondence from numerous envi-ronmental groups and federal and state agencies about a host of issues, from dredging practices to water quality to how to safeguard manatees before using dynamite. In their time, each reflected a critical issue that brought environmentalists to loggerheads with the Corps. At the same time, however, they also represented a new way of doing things around Tampa Bay. Gibbons's intervention signaled that the environmentalists now had a seat at the table in determining the future of the port. He didn't bestow it on them; they had earned it. A good politician, Gibbons under-stood that Tampa residents wanted a busy port, but most were no longer willing to have it at the expense of a healthy bay. He also knew they had the tools they needed—particularly federal and state environmental laws—to challenge anything they felt might harm the bay further.

Environmentalism reshaped the Bay Area's political scene. Old-school champions of growth at any cost didn't go away, but an increasing number of politicians responded to constituents' demands for change and new voices came to the fore. Harold Wilson and Mary Grizzle were among those listening. So, too, was State Representative Roger Wilson, who took up a number of environmental issues, most notably the designation of Pinellas County's remaining publicly owned coast, including large swaths of western Old Tampa Bay, as an aquatic preserve.[26] Across the bay, T. Terrell Sessums, a Davis Islands attorney who rose to become speaker of the Florida House of Representatives, drafted the 1967 Florida Air and Water Pollution Control Act.[27] Tampa's Guy Spicola, who worked with Sessums on that bill, rose to be chair of the House Environmental Protection Committee, where he led the development of regional water policy for the Bay Area.[28]

For all the importance of federal and state law in establishing a framework for saving Tampa Bay, political changes at the local level proved critical in this era. County commissions, city councils, zoning and conservation boards, port authorities, and other local government bodies handled the day-to-day business of governing and often had the final say on the scope and operation of environmental programs. Most members of these entities deferred to the interests of the business community on environmental (and most other) questions. The election to these bodies in the 1970s of people willing to advocate for the health of Tampa Bay signaled how profoundly the political environment shifted in these years.

Tampa Bay found its greatest local political champion in that era. Born in 1936, Jan Kaminis Platt grew up swimming, fishing, and boating in Tampa Bay. In addition to being an expert sailor, she was a brilliant student. After graduating from Florida State University, she tried law school at the University of Florida, but, as the only woman enrolled, found it unwelcoming. She returned to Tampa, taught high school, and became active in the League of Women Voters. She then moved to leadership positions in several community organizations. In addition to leading the Suncoast Girl Scout Council, she headed the Tampa Federation of Garden Clubs and served on the Community Coordinating Council, a volunteer organization designed to propose solutions for various local problems. In 1973, she turned to politics and won a seat on the Tampa City Council.[29]

In this role, she also served on the Tampa Bay Regional Planning Council (TBRPC)—a supragovernmental agency designed to address a host of issues from a regional perspective. She also sat on Governor Reubin Askew's Task Force on Coastal Zone Management.[30] She won a seat on the Hillsborough County Commission in 1978, which she retained until retirement in 2004. Platt was part of a generation of Bay Area women politicians who challenged the good ol' boy system that had run both Tampa and the state for generations. Others included Mary Grizzle, Betty Castor, who preceded Platt on the county commission before election to the Florida Senate, and Mary Figg, who also served as the president of the Hillsborough League of Women Voters before being elected to represent Lutz in the state House of Representatives for more than a decade.

The environment, especially the health of Tampa Bay, was always on the top of Platt's agenda. She chaired the county's Environmental Protection Commission, where she was a staunch defender of Roger Stewart's work. No radical, she instead encouraged various groups with interests in the bay to cooperate in the search for what might be described as practical solutions that recognized the need to balance the demands of shippers, recreational interests, commercial fishermen, homeowners, and others while putting long-term sustainability first. Possessing both personal and public integrity—the *Tampa Times* once described her as "a responsible voice of calm reason in a pond of impulsiveness, irrationality, and political posturing"—she was able to build trust between people who disagreed.[31] Platt's serenity contrasted with the outspokenness of the Bay Area's firebrand environmentalists, but they never doubted which side she was on.

Under Platt's leadership, the county commission made a stronger commitment to protecting Tampa Bay. This could be seen in something as mundane as the development of Upper Tampa Bay Park. The county first took possession of the 435-acre property, which had been part of the Bower Tract, around 1969. Calling it Old Tampa Bay Regional Park, the county's parks department planned to make it a recreation-intensive park. Plans included paved roads connecting a large full-service campground, an equally large picnic area, a marina and boat launch, a fishing pier, and a large swimming beach with a concession stand. The whole project would have required extensive dredging of the bay bottom as well as the destruction of some of the last tidal flats, mangrove forests, freshwater wetlands, and upland pine flatwoods in northern Old Tampa Bay, along with several archaeological sites.[32] Without money to build the park, though, the county sat on the plan for nearly a decade.

In 1978, the county received funds to develop the park but, in keeping with the changing political environment, the old plan was out, replaced by the goal of preserving the tract's natural features. Gone were the camper hookups, the marina, and the beach—replaced instead by nature trails, a nature center, primitive camps sites, and boardwalks.[33] Environmentalists worried that even that much development was too much. Concerned about the impact of roadbuilding in the park, Anita Scharf of the Tampa Audubon Society reminded county commissioners that the goal of the park was to "ensure, not only for ourselves, but for future generations . . . a place where they will be able to enjoy the natural beauties of an undisturbed Florida bay land."[34] John Sharpe, also of the Tampa Audubon Society, was more concise in his recommendations: "The less man made facilities, the better."[35] When the park ultimately opened in 1983, its design reflected the priorities of environmentalists: it contained a single access road designed to avoid sensitive areas (including the nest of a pair of then-endangered bald eagles), hiking trails with minimal buildup, and boardwalks that kept hikers off the shoreline. A canoe launch also replaced the boat ramp entirely. The nature center, meanwhile, instructed visitors about the fragility of the park's ecology. A few picnic pavilions and a single playground were the only concessions to those wanting a traditional park.

Such preservation efforts were not limited to Hillsborough County. In 1974, the state purchased Weedon Island from Florida Power and set out on a multiyear ecological restoration project. Like Upper Tampa Park, the idea was to create space where people might encounter the more-or-less natural bay.[36] In 1980, Pinellas County voters agreed to a slight increase in property taxes to purchase nearly 1,500 acres of shoreline that stretched north from the Gandy Bridge. The Gateway Tract, as it is known, was one of the last contiguous mangrove forests on Old Tampa Bay. Unlike Weedon Island, however, the county purchased it with the idea of setting it aside without making it into a park.[37]

Despite progress on some issues, many problems seemed intractable by the late 1970s and early 1980s. Development continued along the shoreline. Neighborhoods started in the 1950s and the 1960s, like Bay Crest, Tampa Shores, and Apollo Beach grew. At the same time, new residential and commercial properties like Feather Sound opened. Despite federal mandates to make the water cleaner, pollution in some areas grew worse as the stream of new residents—the Bay Area's population nearly doubled between 1970 and 1990—taxed the capacity of even new sewage facilities. In 1981, legislators from Manatee County led the charge to repeal the Wilson-Grizzle Act. New plants could once again return lightly treated sewage to the bay's watershed. (In 1987, Mary Grizzle, then in the Florida Senate, and Florida state representative Mary Figg shepherded through a replacement bill that prevented plants from dumping directly into Tampa Bay.) Fish kills brought on by localized eutrophication events, like one in 1977 that littered the shore of Safety Harbor's Philippe Park with tons of every species the bay offered, occurred regularly enough that they irritated but rarely surprised residents.[38] Phosphorus pollution, too, grew worse, not only because of increased runoff from the phosphate mines east of Tampa but also because of increased runoff from yards and golf courses made green with chemical fertilizers. An accumulation of low-level radioactive runoff from the Gardinier phosphate plant created a dead zone at the mouth of the Alafia River.[39] Tampa Bay's commercial fishery was all but dead.[40] Commercial shrimpers remained but pushed deeper into coastal areas because of the lack of shrimp in open waters. Their trawls damaged seagrass flats from Old Tampa Bay to Cockroach Bay. Meanwhile, siltation vexed Old Tampa Bay's shoreline. In 1981, reporters from *Sports Illustrated* summed it up: "Tampa Bay, once a glory of the state, is . . . filthy."[41]

Environmentalists and their political allies recognized that saving Tampa Bay could not be done by playing Whac-a-Mole with every threat that popped up. Instead, they needed a more systematic approach to managing both the various problems and the shifting coalitions that had stakes in their resolution. They took the first step by literally mapping the bay's ecology and the threats it faced. Hoping to formalize the conversations the bay's scientists were already having, Joseph Simon and Robin Lewis convened experts at USF for the Tampa Bay Area Scientific Information Symposium (BASIS) in May 1982. The program's papers told the story of the bay, from its geology and hydrology to its biology and history. The effects of development and pollution received special attention. While many papers were heavy on numbers and graphs, organizers encouraged the scholars "to make presentations of use to non-technical persons."[42] The goal was not to do science for the sake of science but to do science for the sake of Tampa Bay.

Inspired by the event, Jan Platt requested that the state Department of Environmental Regulation form a commission to study the bay.[43] State Representative Mary Figg sponsored the creation of the Tampa Bay Management Study Commission that same year.[44] In 1985, the commission produced a report, "The Future of Tampa Bay." Robin Lewis, who had by then become an expert on Tampa Bay's seagrasses, was one of the lead authors. The commission identified a long list of issues that needed addressing, from water quality to fishery management to coastal preservation and restoration. Commissioners also proposed a permanent management infrastructure that would link political and economic leaders together with concerned citizens' groups to bring sustained attention to fixing the bay.[45]

The Tampa Bay Regional Planning Council soon after created the Agency on Bay Management (ABM) to take on that very job. The ABM consisted of forty-five members drawn from a variety of government agencies, the business community, outdoor sports organizations, and environmental groups. A full-time professional staff of scientists, aides, and communications specialists carried out the agency's day-to-day work. Chaired from the beginning by Jan Platt, the ABM faced a monumental task. Less than a year into its work, however, the agency's staff identified forty-two specific areas of concern. The ABM had no ability to force change, however. It could simply identify the problems, sponsor additional studies, fund projects like seagrass restoration and beach cleanups, and pressure relevant political bodies to rein in abuses in their jurisdictions. This made for slow going, but the ABM made strides in bringing more attention to the bay's problems, in convincing political leaders to do more to protect the bay, and in providing a framework for what a future bay might look like. The ABM, one staffer remembered, brought "us all together to start communicating. . . . The members of the Agency on Bay Management would put their little domains aside in order to better address the problems of Tampa Bay."[46]

The ABM's activities proved especially helpful when industry set its eyes on the bay's last major undeveloped section: Cockroach Bay. Named for its abundant horseshoe crabs, not the probably equally numerous palmetto bugs, Cockroach Bay is notable for its chain of mangrove islands, one of which, Cockroach Key, has a Safety Harbor–era mound and significant archeological resources. In 1976, the state designated Cockroach Bay as an aquatic preserve and set its boundaries from the southern shore of the Little Manatee River to the line separating Hillsborough and Manatee Counties.[47] In the 1980s, the area boasted clean water, sizable seagrass meadows and salt marshes, and healthy animal populations. It wasn't pristine, but it was about as close as one could get to seeing what Tampa Bay must have looked like a century earlier.

Despite its designation as a preserve, private interests owned the land in and around the bay. Developers, seeing the potential for growth in the Ruskin area, eyed its shoreline. In late 1984, Leisey Shell Pit, Inc. unveiled plans to build what it called Mangrove Bay Marina at an abandoned shell mine it owned in the area known as Little Cockroach Bay, just south of the Little Manatee River. In addition to an 872-slip marina, the company planned to develop 369 acres for housing and other facilities. Building all this would require a lot of dredging. The marina complex itself needed the dredging of upward of 30,000 cubic yards (more than 2,200 dump trucks) of sandy bay bottom and healthy oyster beds. Connecting the marina to Tampa Bay required the construction of a more-than-half-mile-long, eighty-foot-wide channel that would destroy nearly six acres of seagrasses and an acre of mangroves. As the waters of Cockroach Bay were still designated safe for shellfish harvesting, the state Department of Natural Resources denied the company the permit to do the work.[48] The denial of the permit was just the opening act in a drama that lasted more than a year. The company appealed, hoping the governor might overturn the decision. The governor's office sent it back to local agencies, including the ABM, for technical review. Each said no to the permit. The company offered to alter the digging, to build a modern sewage plant, and to mitigate the damage by planting seagrasses elsewhere.[49] Still, the agencies stood firm and issued the final permit denial in June 1987.[50]

The defeat demonstrated the willingness of state and local leaders to use their authority to limit development on the bay. In the 1980s, state and local governments followed a preservationist impulse and established programs to build on the environmental laws of the previous decade. In 1979, the legislature created the Conservation and Recreation Land (CARL) program, which purchased endangered lands. In 1987, it passed the Surface Water Improvement and Management Act (SWIM), which established programs within each of the state's water management districts to protect and improve fresh- and saltwater resources. That same year, Hillsborough County created its own Environmental Lands Acquisition and Protection Program (ELAPP) to purchase sensitive lands. Without a doubt, the company would have received the permit had it applied even a decade earlier. The marina's defeat signaled the changing ethos about balancing nature and economic growth. Supporters of the marina pointed to the jobs it would create in an area in need of them, the convenience it would provide for boaters, and even the benefits of new water infrastructure replacing existing leaky pipes.[51] None, however, could convince political leaders and state regulators that those benefits would be worth the cost of dredging Cockroach Bay or exposing it to the threat of pollution.

The defeat also showed the growing power of Tampa Bay's environmental advocates. News of the planned development moved through the environmentalist grapevine as soon as it broke. Organizations like the Hillsborough County Environmental Coalition and ManaSota-88 held meetings to learn about the development and its impact.[52] Members called and wrote their political leaders in opposition to the plan. Others spoke to their neighbors in the hope of building coalitions to defeat it. "We all take it for granted—we're not educated to the environment—we're trying to protect ourselves from ourselves," said Gus Muench trying to rally a group of Sun City Center residents.[53]

No one worked harder to defend Cockroach Bay than Gus Muench. Muench grew up in Tampa, learned to catch mullet in Papys Bayou, and ended up in Ruskin, where, in addition to his job at the telephone company, he spent every day working crab traps and fishing in Cockroach Bay. He had witnessed the fights to save the bay through the 1970s from a distance. The threat to Cockroach Bay spurred him to action. "I used to let other people do my talkin' for me," he told journalist Jeff Klinkenberg in 1987. "Now I say what I think. Most of us are so involved in just livin' that we don't take time to get involved in the things we should. The result is bad government."[54] Muench joined the Sierra Club and ManaSota-88 and then created the All Release Sport-Fishing Society, the Little Manatee Preservation Committee, and a company devoted to turning seawalls into viable oyster reefs. He wrote letters to political leaders and newspaper editors and showed up at pretty much any meeting about the environment in southern Hillsborough County (and a lot of them in downtown Tampa, too.) He'd take any politician or journalist willing to get in his crab boat all around Cockroach Bay to show it to them from his point of view. He'd cook mullet for volunteers planting seagrasses. He even once took blue crabs from Cockroach Bay and cooked them for politicians in Tallahassee to celebrate Tampa Bay Day.[55] He wasn't afraid to stand up to environmentalists and scientists when he felt they misunderstood the issues as fishermen saw them. And he'd call out the fishermen when their propellers scarred seagrass beds or they failed to obey the law or common sense when it came to fishing.

Muench's greatest hope was to protect Cockroach Bay by having the state buy the property. To Muench, few places better matched CARL's mandate to purchase land of "environmental and cultural significance."[56] In 1986, he started a campaign—and another organization out of his home, "The Committee to Buy the Isles of Cockroach Bay"—to get the area put on CARL's shopping list. (County officials had floated the idea of buying at least Cockroach Key a few years earlier but had prioritized the Bower Tract when money came down from the CARL program.) "The communities surrounding Tampa Bay . . . feel strongly regarding the importance of this pristine aquatic and wilderness area, plus its historical value," he wrote to administrators. "It's imperative to the ecosystem of Tampa Bay that we forever maintain the viable marine and wildlife habitats of Little Cockroach Bay. . . . Let's all of us make one big push to show our concern for the well-being of the bay."[57] A long list of environmental groups from around the Bay Area joined Muench's campaign. Local media publicized and supported his work.[58] In turn, members of the Hillsborough County Commission agreed to support calls for purchasing the bay's islands.[59] CARL administrators listened. Unfortunately, negotiations with the property's owner deadlocked for three years. But on 17 August 1989, Hillsborough County purchased all but one of the islands in Cockroach Bay using ELAPP funds.[60] Muench and environmentalists around Tampa Bay cheered!

Unfortunately, one major threat loomed. About a year before the county completed its purchase, the Tampa Electric Company (TECO) settled on a site for a new plant right along the shore of Cockroach Bay.[61] The company, which opened its Big Bend plant north of Apollo Beach in 1970 and expanded it over the years, argued that the site was necessary to handle future demand. Environmentalists balked at the location. They feared it would undo all they had done and were doing to save Cockroach Bay by introducing pollution both in the construction of the plant and in its operation.[62] "Allowing another industrial plant on the bay's shoreline will seriously damage the estuary, destroy the positive results that have resulted from efforts to restore bay water quality, and adversely affect part of the Cockroach Bay Aquatic Preserve," ManaSota-88 explained, adding that "after their opposition to the marina, approval of by the [Hillsborough County] Commission would be contradictory and extremely damaging to the public."[63]

But fighting TECO was very different from fighting Leisey Shell Pit, Inc. Hillsborough County's sole electricity provider and one of the county's largest employers, TECO had many friends on the Hillsborough County Commission. Commissioners had in fact designated that TECO alone could develop the tract of land in its growth plans for the county. But that did not stop environmental groups from organizing to block construction of the plant. For two years, they held meetings and rallies, pleaded with politicians, and supported studies to demonstrate the potential calamity they believed the plant might bring. TECO responded with an all-out publicity campaign to build support for the plant.

In the fight, though, environmentalists had a powerful ally: the Agency on Bay Management. Almost as soon as TECO announced that it was considering the site, Jan Platt and other leaders at ABM called it a bad idea. "Safeguards need to be taken to protect the water, the sea grasses, the mangroves, and the natural areas," she told journalists, adding that the plant would "upset a vital ecological component of the Tampa Bay estuarine system."[64] At its October 1988 meeting, the ABM's member agencies agreed that Cockroach Bay was "not appropriate for a power generating facility" and for the next year, ABM stood with environmentalists in opposing the plant.[65] When the Hillsborough County Commission voted to support the plant, ABM took the rare step of publicly challenging the commission in hearings before state authorities.[66] Ultimately, opposition grew enough that TECO scrapped plans for the plant. It proved too unpopular with the public, and opposition from political leaders was growing.

In an earlier era, the environmentalists might have been voices in the wilderness when it came to protecting Tampa Bay. By the late 1980s, though, institutions like the ABM provided the space necessary for negotiation. The ABM had little legal power and couldn't solve every problem, but it helped by bringing knowledge of the bay's ecology together with the political heft necessary to build coalitions to defeat things that would further endanger the bay's health. The ABM couldn't give everyone what they wanted. Some in the business community and political circles felt it was too intrusive. Environmentalists sometimes grew frustrated that it could not take more dramatic action to save the bay. But that is the nature of coalitional institutions. Balancing multiple conflicting interests is often a recipe for disaster, but the ABM succeeded because it offered an institutionalized version of connectedness and balance. "One of the things that we had in Tampa Bay, which I think was exceptional" remembered Richard Eckenrod, SWIM coordinator for SWFWMD in the 1980s, "was the situation where the diverse stakeholders in the bay could work together, they were talking to each other from the get go."[67] ABM succeeded because people had figured out that they were all part of the estuary known as Tampa Bay.

In 1987, Congress created the National Estuary Program to highlight the importance of estuarine spaces and provide funds to study, monitor, and improve their health. In 1989, recognizing the opportunities offered by the program, Bay Area political and business leaders, especially those at the ABM, sought to have Tampa Bay designated a national estuary. The following year, Congress added Tampa Bay to the list of estuaries of "national significance," and, the year after that, the Tampa Bay Estuary Program (TBEP) was created.

Following the example of ABM, the TBEP established a series of priorities for protecting and restoring the bay. Those priorities, noted Eckenrod, who went on to become TBEP's first director, reflected the "goal of restoring the balance."[68] Doing this meant first cleaning up the water. On one end of the water cycle, this required cutting down on nutrients and pollution entering the bay. TBEP could do little on its own, but it helped to coordinate the work. In establishing the Nitrogen Management Consortium, for example, it brought together scientists and policymakers to figure out how to turn back the clock on nitrogen levels to 1950s levels. Meetings of this group framed policy for pollution levels and rules for dealing with noncompliant organizations, including local governments. TBEP similarly worked on controlling other pollutants, like phosphorus. It was not always easy, and there were massive setbacks (like a 2004 spill of 65 million gallons of acidic wastewater into the Alafia River), but over time the program has nurtured the sorts of collaborative work necessary to protect Tampa Bay.[69]

The TBEP has similarly nurtured the bay itself. Its 1996 Comprehensive Conservation Management Plan, the goal-setting document required by the EPA, established expanding the size of the bay's seagrass meadows as a top priority. With roughly 24,000 acres left in 1996, the TBEP set a goal of getting the bay back up to 38,500 acres. There was no specific timeline for accomplishing this. Everyone knew it might take decades but

Figure 13. Tampa Bay Estuary Program Repair Kit. Starting in 1991, the Tampa Bay Estuary Program developed educational programs to help Bay Area residents better understand their connections to Tampa Bay. The Tampa Bay Repair Kit, a 20-page illustrated pamphlet published in the early 1990s was designed to educate citizens on how their everyday activities, from water use to fertilizer applications, affected the bay. Its overarching message was the connections between human activities and the ecological health of the bay. At the same time, it highlighted the ability of everyday people to help save Tampa Bay. Credit: Jan Platt Papers, USF Libraries, Special Collections, Tampa.

felt ambition was necessary in the moment. TBEP worked with local governments to find appropriate sites for restoration, sponsored educational outreach programs, and coordinated seagrass planting events. Over time, as water quality improved, the meadows expanded. By 2015, seagrasses covered more bay bottom—more than 40,000 acres—than anyone could remember. Reaching this goal would not have been possible without the support of individuals and organizations all around the Bay Area. Tens of thousands have planted seagrasses and participated in cleanups over the past thirty years.

Peter A. Clark understood the challenges of sponsoring and maintaining these kinds of restoration programs for an agency like the TBEP, which, by design, had a small staff. Clark had worked for the ABM as an environmental planner from its inception and knew the bay's problems

Figure 14. Oyster bag installation, 2016. Volunteers install oyster bags on the shore of Fantasy Island in Hillsborough Bay as part of Tampa Bay Watch's Community Oyster Reef Enhancement Program (CORE). Since 1993, Tampa Bay Watch has worked with the Tampa Bay Estuary Program, local governments and schools, and thousands of volunteers to restore key components of the bay's ecosystem. Photo courtesy of Tampa Bay Watch.

as well as anyone. He felt Tampa Bay needed a regularly staffed, nonprofit organization that could concentrate its efforts on the day-to-day work of cleaning up and restoring the bay. Taking the model of the Hudson's Riverkeeper program, he, with the help of other longtime Bay Area environmentalists, formed Tampa Bay Watch in 1993. Sensing the difficulty of rallying people to act against big problems (that, despite their concreteness, remained abstract in the minds of the public), he decided to focus on hands-on and educational projects that could both teach people about the bay and make them more invested in its health. Tampa Bay Watch sponsored cleanups and seagrass plantings, usually in coordination with the TBEP or local governments. It also worked with local governments to develop management programs, as it did with the City of Tampa for cleaning up McKay Bay. Over time, educational programs became central to Tampa Bay Watch's work. Its most successful program, Bay Grasses in Classes, enlisted students in area schools to grow and transplant seagrasses. It also worked with recreational fishers, both to teach them about the bay and to provide paths so that their knowledge of the water could reach scientists and policymakers. Thirty years on, Tampa Bay Watch has grown significantly and continues the work of making people better aware of the water around them.[70]

Together, the TBEP, Tampa Bay Watch, and the entire framework for managing Tampa Bay have proved a fitting culmination of the work that so many did to make people see that Tampa Bay is not just a port or a playground or a sewer, but an estuary, a place where people and nature meet in a delicate balance. They have made Tampa Bay not only a cleaner place, but a more connected and more balanced one as well.

EPILOGUE

There are times when it turns out that taking the long way is a blessing. When I started this book in 2017, I expected to finish it within two years. But life, including a global pandemic, intervened, and writing it took me longer than I expected. In the end, this has made a better book. In the beginning, I expected to write a mostly upbeat story of how dedicated public servants and energized citizens came together to fix an ecosystem once thought beyond repair. It was, I told friends, a story with a happy ending, not something we often get to tell in the field of environmental history. That is still largely my assessment, too. Tampa Bay is cleaner than it was when I was growing up there from the 1970s to the 1990s. (Disclosure: it was my neighborhood that Mike Wallace toured with Roger Stewart and James Shimberg for his 1974 *60 Minutes* report.) And today there are much stronger guardrails in place to protect it. When videos of cownose rays schooling in Boca Ciega Bay or mating manatees stopping traffic on the Courtney Campbell Causeway pop up on my social media feeds, I'm reminded of the people who worked so hard to make the bay cleaner. And when I see updates from the Tampa Bay Estuary Program, Tampa Bay Watch, and other groups about their efforts to plant seagrasses, seed oyster reefs, protect seabirds, and teach people about the beauty and value of Tampa Bay, I am thankful for their work and heartened for the future.

Events of recent years, ones I would have missed had I written faster, have tempered my optimism. The most explosive of these was the release of 215 million gallons of phosphorus-laden water into the bay between Cockroach Bay and Terra Ceia Bay in March 2021. The liner in a reservoir at the site of a former phosphate processing plant on Piney Point sprang a leak, placing massive pressure on one of the reservoir's walls. The reservoir held more than a half billion gallons of rainwater contaminated with phosphorus and a slew of other chemicals from the phosphogypsum stack created by the plant and left behind when it closed. If the reservoir's wall broke, the water would inundate the surrounding area, risking the lives and property of more than three hundred residents as well as the inmates at the Manatee County Jail. Officials at the Florida Department of Environmental Protection decided to reduce pressure on the wall by pumping water out of the reservoir and into Tampa Bay.[1]

Environmentalists were terrified about what it might mean for the bay. Not only was the water loaded with phosphorus, it contained the heavy metal cadmium and had a pH of 4.6, only slightly more acidic than coffee, but acidic enough to stress fish, especially those used to Tampa Bay's slightly basic waters.[2] Scientists worried it might fuel an algal bloom that could lead to eutrophication or a red tide bloom. Both could cause massive fish kills. Tampa Bay Watch and other environmentalists rushed to keep boats on the water to monitor the effects of the pumping. Most agreed the pumping was the least bad option facing the state in this situation, but they weren't happy. The state stopped pumping within weeks, but the effects lingered. By summer, residents' fears had been realized: the release fed a red tide event that killed more than 1,600 tons of marine life in the bay's lower reaches.[3] And the effects continue to linger.

Piney Point is a reminder that any attempt to put a happy ending on the story of Tampa Bay is premature. From aging phosphogypsum stacks to leaky sewage lines, legacies of the past haunt the Bay Area and threaten the delicate balance years of policymaking and hard work have sustained. Dealing with them will be difficult. Solutions are expensive and certain to engender political strife. While we can hope present and future generations will continue the spirit of cooperation that fueled the turnaround in the 1980s and 1990s, it is not guaranteed.

Tampa Bay's waterfront has been radically transformed in recent years. From the Tampa Riverwalk and Water Street projects to the new St. Pete Pier, Tampa Bay is in the midst of a renaissance of public recreational access. Happily, new designs for all three main spans across the bay have taken the ecological needs of the bay into account. They even hire spotters to look out for manatees these days! But growth will continue to present a challenge. The Tampa Bay Metro Area, home to nearly 4 million people, is the nation's fifteenth largest.[4] These people will tax the region's resources and place more strain on existing systems designed to mitigate pollution. Already, scientists have noted that seagrass acreage, once the great benchmark of the bay's comeback, has declined consistently in recent years and is down 30 percent since peaking in 2016. Increased fertilizer runoff, scientists suspect, is the cause.[5] While localities have moved to ban summer fertilizer applications, only time will show if runoff can be curtailed and the seagrass meadows saved. And even if people break from their desire for green lawns, demand for things like waterfront housing will continue and likely increase. As recently as 2018, developers sought to fill bay bottom on Rocky Point to expand a business park. It is fully within the realm of possibility that pressure to open protected areas to development will increase if citizens are not vigilant.

Protecting Tampa Bay will not be solely a local matter in the future, though. Global issues will also play a role. Low slung and lined with development, Tampa Bay will feel the effects of sea level rise over the coming decades. A 2017 *Washington Post* story by environmental journalist Darryl Fears explored this very problem and concluded that rising sea levels make Tampa Bay one of the most vulnerable places in the United States for a storm to cause catastrophic damage.[6] Save that last minute turn, Hurricane Ian might have proved him right. How long can the Bay Area's luck hold? Beyond the threat of storm surges, sea level rise will require greater efforts than already expended to make the bay's coastline resilient. Meanwhile, climate change more broadly, with its increased water temperatures and acidification, are going to present challenges to the bay's flora and fauna. Microplastics, too, have been found in Tampa Bay in recent years. The product of plastic pollution breaking down in salt water, microplastics pollute oceans the world over, killing fish and seabirds and adding numerous chemicals to the food webs. Their effect on Tampa Bay is not yet known, but it is likely to be great over the next century. No movement to preserve Tampa Bay will succeed without considering the global alongside the local.

In 1985, my fifth grade class took a field trip to the Courtney Campbell Causeway. We pulled what seemed like a large dragnet in from maybe twenty yards out to see what we would find. (Keep your shoes on and watch for holes! our teacher warned us.) The highlight of our catch? A single pufferfish, angry that we had snared it, and a small stingray. Hardly a scientific sample but suggestive of just how bad things were back then. Thanks to the work of so many, Tampa Bay is far healthier today. We have come to understand the importance of protecting the bay as an estuary and have been willing to limit our impacts on it. Because of that, I have confidence that, nearly forty years after my disappointing day at the beach, another fifth grader might find much more in her net. The challenge we face is what her children and grandchildren may find in theirs.

ACKNOWLEDGMENTS

I first want to thank Andrew Frank for approaching me to see if I had any ideas for the Florida in Focus book series. I'd had an idea for the environmental history of Tampa Bay churning in the back of my mind for some time, and I am so glad I got the opportunity to write it. Thanks also to Sian Hunter at the University Press of Florida for guiding this project. Her encouragement and patience have helped so much along the way.

My colleagues at Florida Atlantic University have been a constant well of support for this project. Sandy Norman, Steve Engle, Eric Hanne, Doug Kanter, Patty Kollander, Ben Lowe, and all the rest of you: thank you for all the times I talked your ear off about this book. Thanks also to Dean Michael Horswell and the Dorothy F. Schmidt College of Arts and Letters for providing funds to help support the research for this book and the creation of the map.

Thanks to Andy Huse at Special Collections, University of South Florida Libraries, Tampa, especially for the tip to look at the Skip Gandy photos. I appreciate everyone in special collections for keeping those (many, many) boxes of Jan Platt's papers coming when I kept asking for more.

When I started this project, I had no idea of the community of scholars and activists who had been thinking about Tampa Bay, too. Thanks to Andy Fairbanks for introducing me to that world. Thanks to Ann Paul for inviting me to help on her excellent exhibition at the Tampa Bay History Center, *Conservation: A Bird's Eye View*.

Though neither has seen a word of this book yet, I need to thank Ray Arsenault and Gary Mormino for teaching me that writing about Florida history is important. I hope you both like it.

And thank you Rachel and Charlie. For everything.

NOTES

Introduction

1 "Potential Historic Catastrophe" and "'Get Out Right Now': Evacuations Underway," *Tampa Bay Times*, 27 September 2022.
2 "A New Report Ranks Ian as the Third-Costliest Hurricane on Record in the U.S.," WFSU Public Media, 11 January 2023. https://news.wfsu.org/state-news/2023–01–11/a-new-report-ranks-ian-as-the-third-costliest-hurricane-on-record-in-the-u-s.

Chapter 1. Larder

1 Jerald T. Milanich, *Florida's Indians from Ancient Times to the Present* (Gainesville: University Press of Florida, 1998), 3–5. "Was Hillsborough Bay Once a Great Freshwater Lake?" *Tampa Times*, 22 July 1921. Gerold Morrison and Kimberly K. Yates, "Origin and Evolution of Tampa Bay," chapter 3 in *Integrating Science and Resource Management in Tampa Bay, Florida*, U.S. Geological Survey Circular 1348, ed. Kimberly K. Yates, Holly Greening, and Gerold Morrison (Washington, D.C.: U.S. Department of the Interior, 2011), 38–45.
2 Albert C. Goodyear, Sam B. Upchurch, Mark J. Brooks, and Nancy N. Goodyear, "Paleo-Indian Manifestations in the Tampa Bay Region, Florida," *Florida Anthropologist* 36, nos. 1–2 (March–June 1983): 40–66.
3 Morrison and Yates, "Origin and Evolution of Tampa Bay," 38–41. Roy R. Lewis III and Ernest D. Estevez, *The Ecology of Tampa Bay, Florida: An Estuarine Profile* (Washington, D.C.: U.S. Fish and Wildlife Service, 1988), 11–12.
4 Lyman O. Warren, William Thompson, and Ripley P. Bullen, "The Culbreath Bayou Site, Hillsborough County, Florida," *Florida Anthropologist* 20, nos. 3–4 (September–December 1967): 146–63. Milanich, *Florida's Indians from Ancient Times to the Present*, 34–36.
5 *Yat Kitischee: The Archaeology of a Prehistoric Coastal Hamlet*, a report by Janus Research for the Board of County Commissioners of Pinellas County, Florida (Clearwater: Pinellas County Department of Environmental Management, 2002), 18–22.
6 Robert J. Austin, Jeffrey M. Mitchem, and Brent R. Weisman, "Radiocarbon Dates and the Late Prehistory of Tampa Bay," in *New Histories of Pre-Columbian Florida*, ed. Neill J. Wallis and Asa R. Randall (Gainesville: University Press of Florida, 2014), 101–3.

7 Milanich, *Florida's Indians from Ancient Times to the Present*, 102–3.
8 Milanich, *Archaeology of Precolumbian Florida* (Gainesville: University Press of Florida, 1994), 159–164, 207–41.
9 Robin C. Brown, *Florida's First People: 12,000 Years of Human History* (Sarasota: Pineapple, 2013), 152–53.
10 Milanich, *Florida's Indians from Ancient Times to the Present*, 103–4.
11 John W. Griffin and Ripley P. Bullen, *The Safety Harbor Site–Pinellas County Florida*, Florida Anthropological Society Publications, no. 2 (Gainesville: University of Florida, 1950), 29.
12 Griffin and Bullen, *The Safety Harbor Site–Pinellas County Florida*, 29.
13 Griffin and Bullen, *The Safety Harbor Site–Pinellas County Florida*, 29. Scott Cave, "Madalena: The Entangled History of One Indigenous Floridian Woman in the Atlantic World," *The Americas* 74, no. 2 (April 2017): 178. Jeffrey McClain Mitchem, "Redefining Safety Harbor: Late Prehistoric/Protohistoric Archaeology in West Peninsular Florida" (PhD diss., University of Florida, 1989), 50–57.
14 John H. Hann, *Indians of Central and South Florida, 1513–1763* (Gainesville: University Press of Florida, 2003), 113.
15 Milanich, *Florida's Indians from Ancient Times to the Present*, 103–4.
16 Milanich, *Florida's Indians from Ancient Times to the Present*, 103.
17 Milanich, *Florida's Indians from Ancient Times to the Present*, 107. Jeffrey M. Mitchem, "Safety Harbor: Mississippian Influence in the Circum-Tampa Bay Region," in *Late Prehistoric Florida: Archaeology at the Edge of the Mississippian World*, ed. Keith Ashley and Nancy Marie White (Gainesville: University Press of Florida, 2012), 172–85.

Chapter 2. Borderland

1 Samuel Turner, "Juan Ponce de León and the Discovery of Florida Reconsidered," *Florida Historical Quarterly* 92, no. 1 (Summer 2013): 5.
2 John E. Worth, ed. and trans., *Discovering Florida: First-Contact Narratives from Spanish Expeditions along the Lower Gulf Coast* (Gainesville: University Press of Florida, 2014), 18.
3 Jerald T. Milanich, *Florida Indians and the Invasion from Europe* (Gainesville: University Press of Florida, 1995), 110.
4 This account of the Narváez expedition is based on Paul E. Hoffman, "Narváez and Cabeza de Vaca in Florida," in *The Forgotten Centuries: Indians and Europeans in the American South, 1521–1704*, ed. Charles Hudson and Carmen Chaves Tesser (Athens: University of Georgia Press, 1994), 50–58.
5 Rolena Adorno and Patrick Charles Pautz, *Álvar Núñez Cabeza de Vaca: His Account, His Life, and the Expedition of Pánfilo de Narváez*, vol. 2 (Lincoln: University of Nebraska Press, 1999), 89–90. Robert S. Weddle, *Spanish Sea: The Gulf of Mexico in North American Discovery* (College Station: Texas A&M University, 1985), 188.
6 Charles Hudson, "The Hernando de Soto Expedition, 1539–1543," in *The Forgotten Centuries: Indians and Europeans in the American South, 1521–1704*, ed. Charles Hudson and Carmen Chaves Tesser (Athens: University of Georgia Press, 1994), 75.

7 While it is not clear that these particular hogs are the ancestors of Florida's modern wild hogs—as is often claimed—they nevertheless grew in number and were part of the introduction of non-native species that was part and parcel of Spanish colonialism. See Abraham Gibson, *Feral Animals in the American South: An Evolutionary History* (Cambridge: Cambridge University Press, 2016), 30, 46.

8 Charles Hudson, *Knights of Spain, Warriors of the Sun: Hernando de Soto and the South's Ancient Chiefdoms* (Athens: University of Georgia Press, 1997), 72–78.

9 Hudson, *Knights of Spain, Warriors of the Sun*, 62–64, 71.

10 Hudson, *Knights of Spain, Warriors of the Sun*, 66.

11 Hudson, *Knights of Spain, Warriors of the Sun*, 85.

12 Hudson, *Knights of Spain, Warriors of the Sun*, 87.

13 Scott Cave, "Madalena: The Entangled History of One Indigenous Floridian Woman in the Atlantic World," *The Americas* 74, no. 2 (April 2017): 171–200.

14 Worth, *Discovering Florida*, 261.

15 Worth, *Discovering Florida*, 262.

16 Worth, *Discovering Florida*, 263–264.

17 John H. Hann, ed. and trans., *Missions to the Calusa*, Ripley P. Bullen series (Gainesville: University Presses of Florida, 1991), 253.

18 Hann, *Missions to the Calusa*, 254.

19 Paul E. Hoffman, *Florida's Frontiers* (Bloomington: Indiana University Press, 2002), 103. Jerald T. Milanich and Charles Hudson, *Hernando de Soto and the Indians of Florida* (Gainesville: University Press of Florida, 1993), 46.

20 Hoffman, *Florida's Frontiers*, 103. Hann, *Missions to the Calusa*, 347–48, 351.

21 Robert S. Weddle, trans., "The Enriquez Barroto Diary," in *La Salle, the Mississippi, and the Gulf*, ed. Robert S. Weddle (College Station: Texas A&M University Press, 1987), 198.

22 John H. Hann, "Demise of the Pojoy and Bomto," *Florida Historical Quarterly* 74 (Fall 1995): 188.

23 They were not the only non-Apalachee remnant group seeking shelter in Apalachee in the late seventeenth century. It seems, too, that non-Tocobaga peoples lived in the towns the Spanish identified as Tocobaga towns in Apalachee. Hann, *Missions to the Calusa*, 347, 349.

24 Hann, *Missions to the Calusa*, 350. In 1718, the Pojoy carried out attacks on the people of Tocobaga in Apalachee, suggesting the rift between the two remained.

25 Ronald Wayne Childers, "Historic Notes and Documents: A Late Seventeenth-Century Journey to Tampa Bay," *Florida Historical Quarterly* 80 (Spring 2002): 504–22.

26 Hann, "Demise of the Pojoy and Bomto," 184–200.

27 Jack D. L. Holmes and John D. Ware, "Juan Baptista Franco and Tampa Bay, 1756," *Tequesta* 28 (1968): 91–98, quotations on 95, 96.

28 John D. Ware, trans., "From Havana to the Port of Tampa, Year of 1757: A Journal of Surveys, Atlantic Ocean, Northern Part" (n.p.), 40. Available at University of South Florida (USF) Libraries, Special Collections, Tampa.

29 Kathryn E. Holland Braund, *Deerskins and Duffels: Creek Indian Trade with Anglo-America, 1685–1815* (Lincoln: University of Nebraska Press, 1996), 62–63.

30 George Gauld, *Observations on the Florida Kays, Reef and Gulf*, London: 1796, n.p.

31 John D. Ware, *George Gauld: Surveyor and Cartographer of the Gulf Coast* (Gainesville: University Presses of Florida, 1982), 54.

32 Holland Braund, *Deerskins and Duffels*, 62.

33 James W. Covington, "Trade Relations between Southwestern Florida and Cuba, 1600–1840," *Florida Historical Quarterly* 38 (Oct. 1959): 117–19. Michelle Honora Zacks, "From Table to Trash: The Rise and Fall of Mullet Fishing in Southwest Florida" (PhD diss., University of Hawaii, 2013), 28–29. James Grant Forbes, *Sketches Historical and Topographical of the Floridas, More Particularly of East Florida* (New York: C. S. Van Winkle, 1821), 118.

34 John E. Worth, "Creolization in Southwest Florida: Cuban Fishers and 'Spanish Indians' ca. 1766–1841," *Historical Archaeology* 46, no. 1 (2012): 148–49, 154.

35 Kenneth W. Porter, *The Black Seminoles: History of a Freedom-Seeking People*, ed. Alcione M. Amos and Thomas P. Senter (Gainesville: University Press of Florida, 1996).

36 Hoffman, *Florida's Frontiers*, 263–65.

37 Brent Richards Weisman, *Unconquered People: Florida's Seminole and Miccosukee Indians* (Gainesville: University Press of Florida, 1999), 44–45.

38 James Gadsden, "The Defences of the Floridas (1818)," *Florida Historical Quarterly* 15, no. 4 (April 1937): 248.

39 American State Papers, Indian Affairs, vol. 2, 1822. In American State Papers: Documents, Legislative and Executive of the Congress of the United States, from the First Session of the Fourteenth to the Second Session of the Nineteenth Congress, Commencing December 4, 1815, and Ending March 3, 1827 (Washington, D.C.: Gales and Seaton, 1834), https://memory.loc.gov/cgi-bin/query/r?ammem/hlaw:@field (DOCID+@lit(sp0082).

40 Canter Brown Jr., "The 'Sarrazota, or Runaway Negro Plantations': Tampa Bay's First Black Community, 1812–1821," *Tampa Bay History* 12, no. 2 (Fall/Winter 1990): 5–19.

41 Rosalyn Howard, "'Looking for Angola': An Archeological and Ethnohistorical Search for a Nineteenth Century Florida Maroon Community and Its Caribbean Connections," *Florida Historical Quarterly* 92, no. 1 (Summer 2013): 35.

42 Brown Jr. "The 'Sarrazota, or Runaway Negro Plantations,'" 5–19, quotations on 13.

43 Brown Jr. "The 'Sarrazota, or Runaway Negro Plantations,'" 5–19, quotations on 13.

44 "An Act for the Protection of the Fisheries on the Coast of Florida, and to Raise Revenue Therefrom," *Compilation of the Public Acts of the Legislative Council of the Territory of Florida Passed Prior to 1840*, comp. John P. Duval (Tallahassee: Samuel S. Sibley, 1839), 375–81, quotation on 380. The effect of the law is hard to measure, and no doubt it was hard to enforce, but evidence from south of Tampa Bay suggests it made conditions more difficult for Cuban fishers. See W. A. Whitehead to Joseph M. White, 22 March 1832, reprinted in U.S. House of Representatives, "Resolution and Act of the Legislative Council of the Territory of Florida for the Protection of the Fisheries of the Said Territory," doc. 201, 9 April 1832, 1.

45 Dorothy Dodd, "Captain Bunce's Tampa Bay Fisheries, 1835–1840," *Florida Historical Quarterly* 25, no. 3 (January 1947): 246–56. James W. Covington, "Trade Relations between Southwestern Florida and Cuba, 1600–1840," *Florida Historical Quarterly* 38 (Oct. 1959): 123–24.

46 Covington, "Trade Relations," 123–24.

47 Charles C. Coe, *Red Patriots: The Story of the Seminoles* (1898; reprint Gainesville: University Presses of Florida, 1974), 160–61.

48 Grant Foreman, *Indian Removal: The Emigration of the Five Civilized Tribes of Indians* (1932; reprint Norman: University of Oklahoma Press, 1957), 362.

49 Letter from commanding officer, Fort Brooke to adjutant general of the army, dated 31 August 1840, reprinted in John T. Sprague, *The Origin, Progress, and Conclusion of the Florida War* (1848), reprinted with an introduction by John K. Mahon (Tampa: University of Tampa Press, 2000), 321–24.

Chapter 3. Port

1 *Rhode Island American*, 6 March 1832.

2 George Ballentine, *Autobiography of an English Soldier in the United States Army: Comprising Observations and Adventures in the States and Mexico* (New York: Stringer & Townsend, 1853), 121.

3 Ballentine, *Autobiography of an English Soldier*, 100–101

4 James B. Kenyon, "Land Use Admixture in the Built-Up Urban Waterfront: Extent and Implications," *Economic Geography* 44 (April 1968): 155.

5 Anthony P. Pizzo, *Tampa Town, 1824–1886: Cracker Village with a Latin Accent* (Miami: Hurricane House, 1968), 1–3, 14.

6 John Solomon Otto, "Hillsborough County (1850): A Community in the South Florida Flatwoods," *Florida Historical Quarterly* 62, no. 2 (October 1983): 183, 185. In 1840, census takers counted about 700 people, most of them soldiers at Fort Brooke. By 1850, more than 2,300 people lived in Hillsborough County, which then included a sizable portion of Central Florida, including all of modern-day Pinellas County.

7 Julia Floyd Smith, *Slavery and Plantation Growth in Antebellum Florida, 1821–1860* (Gainesville: University of Florida Press, 1973), 130–32. Canter Brown Jr., *African Americans on the Tampa Bay Frontier* (Tampa: Tampa Bay History Center, 1997), 23, 37.

8 Raymond Arsenault, *St. Petersburg and the Florida Dream, 1888–1950* (1988; reprint Gainesville: University Press of Florida, 1996), 31–32.

9 *Burlington Free Press*, 10 June 1836.

10 Otto, "Hillsborough County (1850)," 189–90.

11 Lucy D. Jones, "History of Hooker's Point, Port of Tampa, Florida," 2003, Tony Pizzo Papers, box 46, folder 6, USF Libraries, Special Collections, Tampa.

12 *Florida Peninsular*, 28 July 1860, quoted in *Marine History of Tampa Bay, Florida: 19th Century*, p. 31, comp. Julius J. Gordon, 1994, Tampa-Hillsborough County Public Library, Special Collections. For a description of McKay's ship, *Salvor*, see "Herald Marine Correspondence," *New York Herald*, 25 August 1860.

13 *Florida Peninsular*, 3 December 1859, quoted in *Marine History of Tampa Bay*, 28.

14 *Florida Peninsular*, 3 December 1859, quoted in *Marine History of Tampa Bay*, 28, see also 49.

15 Donald H. Thompson and Carol Thompson, *Egmont Key: A History* (Charleston: History Press, 2012), ebook n.p., chapter 3.

16 *Tahoma I*, Dictionary of American Naval Fighting Ships, Naval History and Heritage Command, https://www.history.navy.mil/research/histories/ship-histories/danfs/t/tahoma-i.html, accessed 19 June 2020.

17 "Eastern News," *Idaho Statesman*, 3 January 1865. *Daily Ohio Statesman*, 19 December 1864.

18 "West Coast of Florida," *New York Times*, 17 November 1862.

19 United States Department of the Interior, Census Office, *Statistics of the Population of the United States at the Tenth Census* (Washington, D.C.: Government Printing Office, 1882), 384.

20 "The Land of Flowers," *Troy Messenger*, 4 March 1880.

21 "D.B. McKay Recalls the Days When Schooners Were Tampa's Main Link with the Outside World," *Tampa Times*, 23 December 1935.

22 "Facts About Florida," *Hartford Daily Courant*, 30 November 1875.

23 "The Land of Flowers," *Troy Messenger*, 4 March 1880.

24 George E. Buker, *Sun, Sand, and Water: A History of the Jacksonville District of the U.S. Army Corps of Engineers, 1821–1975* (Washington, D.C.: Government Printing Office, 1981), 133–34.

25 Jack E. Davis, "Alligators and Plume Birds: The Despoliation of Florida's Living Aesthetic," in *Paradise Lost?: The Environmental History of Florida*, ed. Jack E. Davis and Raymond Arsenault (Gainesville: University Press of Florida, 2005), 235–59.

26 W.E.D. Scott, "The Present Condition of Some of the Bird Rookeries of the Gulf Coast of Florida, Third Paper," *The Auk* 4, no. 4 (July 1887): 278, 280.

27 Scott, "The Present Condition," 281–82.

28 "News at St. Petersburg," *Tampa Morning Tribune*, 17 July 1900.

29 "The Letter Box," *St. Nicholas: An Illustrated Magazine for Young Folks* 22, no. 2 (December 1894): 172.

30 1880 and 1910 Federal Census Reports. These figures were calculated from Hillsborough County's population only. Hillsborough County included all of modern-day Pinellas County until 1912. Population growth by decade was as follows: 1880–1890: 157%; 1890–1900: 141%; 1900–1910: 118%.

31 Arsenault, *St. Petersburg and the Florida Dream*, 61–62.

32 "What to Eat," *Atlanta Constitution*, 14 October 1888.

33 "The County's Fish Interests," *Braidentown News*, 12 November 1897. For an early history of Hunter's Point and Cortez, see Ben Green, *Finest Kind: A Celebration of a Florida Fishing Village* (Macon, Ga.: Mercer University Press, 1985), 43–63.

34 "Work on the Wharves," *Tampa Weekly Tribune*, 30 September 1897.

35 "McIlvane Mad," *Tampa Tribune*, 13 August 1895.

36 "The County's Fish Interests," *Braidentown News*, 12 November 1897.

37 Quoted in *The Pensacola Journal*, 20 January 1905.

38 Charles E. Dawson, *A Survey of the Tampa Bay Area*, Florida State Board of Con-

servation Technical Series no. 8 (Florida State Board of Conservation: Tallahassee, 1953), 32.

39 "Oysters and Fish," *Tampa Daily Journal*, 21 March 1890. Gary R. Mormino, "When Oysters Ruled," *Tampa Tribune*, 1 January 2006.

40 Buker, *Sun, Sand, and Water*, 135.

41 "A Flying Trip to Cuba," *New York Times*, 14 July 1889.

42 Mariola Espinosa, *Epidemic Invasions: Yellow Fever and the Limits of Cuban Independence, 1878–1930* (Chicago: University of Chicago Press, 2009), 17.

43 Eirlys Mair Barker, "Seasons of Pestilence: Tampa and Yellow Fever, 1824–1905," (MA thesis, University of South Florida, 1984).

44 Stephen T. Messenger, "The Trains Stop at Tampa: Port Mobilization during the Spanish-American War and the Evolution of Army Deployment Operations," *Army History* 104 (Summer 2017): 34–55.

45 Buker, *Sand, Sun, and Water*, 136. "Man Behind the Port," *Tampa Tribune*, 22 June 1985.

46 Buker, *Sand, Sun, and Water*, 136–40.

47 "$727,500 for Tampa Harbor," *Wall Street Journal*, 8 September 1933.

48 "Great Estuary Fill Is Ready After 12 Years for Industrial Center," *Tampa Daily Times*, 18 October 1924.

49 "Preliminary Examinations–Tampa and Hillsboro Bays," 1927, box 3324, folder 75, Records of the Office of the Chief of Engineers, RG 77, National Archives and Records Administration, Atlanta, Georgia.

50 "Fine Havanas Made in Tampa," *Wall Street Journal*, 18 December 1933.

51 "Drama in Ships Adds Romance to Tampa's Waterfront Activity," *Tampa Tribune*, 11 August 1935.

52 Tampa Customs Collector to Warner P. Sutton, 15 July 1885, Records of the Customs Houses in the South Atlantic States, Tampa, Letters Sent, 1886–1900, box 1, letterbook 1, Records of the Customs Service, RG 36, National Archives and Records Administration, Atlanta, Georgia.

53 L. N. Dantzler to U.S. Army Corps of Engineers, 8 March 1930," box 3188, Records of the Office of the Chief of Engineers, Jacksonville District, RG 77, National Archives, Atlanta, Georgia.

54 U.S Army Corps of Engineers and U.S. Shipping Board, *Port Series No. 8: The Ports of Jacksonville, Miami, Key West, Tampa, and South Boca Grande, Florida*, Part 2: *Ports of Miami and Tampa, Florida* (Washington, D.C.: Government Printing Office, 1931), 90–106.

55 "Channel Plan Approved," *Wall Street Journal*, 6 August 1929.

56 Lewis N. Wynne, "Shipbuilding in Tampa during World War II," *Sunland Tribune* 16 (1990), https://scholarcommons.usf.edu/sunlandtribune/vol16/iss1/13.

57 Jones, "History of Hooker's Point," 2003, Tony Pizzo Papers, box 46, folder 6, USF Libraries, Special Collections, Tampa.

58 "First Shrimp Boat Docks With Catch," *Tampa Tribune*, 7 December 1951.

59 "Tampa Shrimpers Fish for 'Pink Gold,'" *Tampa Tribune*, 14 June 1970.

60 Jones, "History of Hooker's Point," 2003, Tony Pizzo Papers, box 46, folder 6, USF Libraries, Special Collections, Tampa.

61 Tampa Port Authority, *Annual Report, 1967*, Sam M. Gibbons Papers, box 48, folder 1, USF Libraries, Special Collections, Tampa.

62 "Tampa May Invest $5,000,000 in Port," *Tampa Morning Tribune*, 6 January 1926.

63 "Phosphate Hassle in Tampa Bay," *Florida Trend*, July 1965.

64 "Statement of Congressman Sam M. Gibbons before the Public Works Subcommittee," 4 May 1966, Sam M. Gibbons Papers, box 28, folder 4, USF Libraries, Special Collections, Tampa.

65 "Engineers Urge Port Work," *Tampa Tribune*, 14 June 1969.

66 "Deep Channel a Harbor Must," *Tampa Tribune*, 14 June 1970.

67 "Memorandum: Further Improvement of Channels in Tampa Bay and Harbor, Florida," 22 January 1946, U.S. Army Corps of Engineers Records, Jacksonville District, box 62, folder J-827, National Archives, East Point, Georgia.

68 Advertisement, *Tampa Tribune*, 14 June 1970.

Chapter 4. Playground

1 *Photographic Views of the Tampa Bay Hotel* (Buffalo, N.Y.: Matthews-Northrup, 1895).

2 *Photographic Views of the Tampa Bay Hotel.*

3 "The Silver King," *Louisville Courier-Journal*, 3 June 1888.

4 Kevin Kokomoor, "'In the Land of the Tarpon': The Silver King, Sport, and the Development of Southwest Florida, 1885–1915," *Journal of the Gilded Age and Progressive Era* 11, no. 2 (April 2012): 191–224.

5 "From Florida," *Grand Forks Daily Herald*, 6 March 1898. Straub later went on to become owner and editor of the *St. Petersburg Times* and a champion of protecting the bay from overdevelopment.

6 James A. Henshall, *Favorite Fish and Fishing* (New York: Outing Publishing, 1908), 147.

7 George S. Gandy, oral interview by Ann B. Hodgson, 2015, USF Libraries, Oral History Program (OHP), Tampa. In *The Tampa Bay Estuary: An Oral History of Community Collaboration to Restore Ecological Integrity*, audio file T43_00017, https://digitalcommons.usf.edu/tb_estuary_ohp/17.

8 "Powell Tells of the Unfolding of Florida," *Tampa Tribune*, 18 February 1918.

9 "Crabbing Party," *Tampa Tribune*, 5 October 1901.

10 "Picnic Island," *Tampa Journal*, 16 April 1891.

11 Pattie Dervaes, "Ballast Point: Cow Pastures to Condominiums," *Sunland Tribune* 9, no. 1 (1983):27–32, https://scholarcommons.usf.edu/sunlandtribune/vol9/iss1/10.

12 "DeSoto Park Public Place," *Tampa Times*, 28 March 1913.

13 "Preparing for Pleasure," *Tampa Weekly Tribune*, 6 August 1896.

14 "Comparisons: Two Florida Towns and Their Characteristics Discussed by Mr. Straub," *Grand Forks Daily Herald*, 10 April 1898.

15 "About five hundred negroes from this city," *Tampa Tribune*, 4 April 1899.

16 "DeSoto Park Public Place," *Tampa Times*, 28 March 1913.

17 Raymond O. Arsenault, *St. Petersburg and the Florida Dream, 1888–1950* (1988; reprint Gainesville: University Press of Florida, 1996), 87–88.

18 "What Is the W.T.I.A. and What Does It Do?" *St. Petersburg Times*, 2 December 1917.

19 St. Pete Pier History, https://stpetepier.org/history/.

20 "Repression Is a Theme of City's Black History," *St. Petersburg Times*, 19 March 1979.

21 "Seek Bathing Place for Negro Swimmers," *St. Petersburg Times*, 19 June 1923.

22 "South Siders Fight Negro Bathing Beach," *Tampa Tribune*, 27 June 1923.

23 "Extensive Developments in North Shore Property," *St. Petersburg Times*, 7 May 1912.

24 Charles A. Brown, "Bayshore Boulevard: A Brief Look at Its Historic Past," Tony Pizzo Papers, box 5, folder 7, USF Libraries, Special Collections, Tampa. Suburb Beautiful advertisement, *Tampa Tribune*, 8 May 1910.

25 "Sunset Bathing Beach Growing," *Tampa Times*, 19 May 1916.

26 Sunset Park advertisement, *Tampa Tribune*, 30 April 1916. "Lots Sell Fast in Sunset Park," *Tampa Times*, 1 August 1925.

27 "Extensive Developments in North Shore Property," *St. Petersburg Times*, 7 May 1912.

28 Suburb Beautiful advertisement, *Tampa Tribune*, 8 May 1910.

29 Suburb Beautiful advertisement, *Tampa Tribune*, 1 November 1909.

30 Sunset Park advertisement, *Tampa Tribune*, 30 April 1916.

31 Jay Barnes, *Florida's Hurricane History*, 2nd ed. (Chapel Hill: University of North Carolina Press, 2007), 103–8.

32 Arsenault, *St. Petersburg and the Florida Dream*, 198.

33 "Residents of Beach Park Find Fishing Good in Tampa Bay," *Tampa Tribune*, 13 June 1926.

34 Davis Islands advertisement, *Philadelphia Inquirer*, 11 December 1927.

35 Christopher Knowlton, *Bubble in the Sun: The Florida Boom of the 1920s and How It Brought On the Great Depression* (New York: Simon & Schuster, 2020), 117–21. Davis Islands advertisement, Tony Pizzo Papers, box 29, folder 6, USF Libraries, Special Collections, Tampa.

36 "Proposed Yacht Course in Tampa Bay," *Tampa Tribune*, 11 January 1925. It's not entirely clear if he followed through on his plan. Davis died at sea on his way to Europe in 1926.

37 "Little Wind and Tide Make Tampa Bay Ideal," *Tampa Tribune*, 11 January 1925.

38 Shore Acres advertisement, *St. Petersburg Times*, 25 March 1923.

39 "Working on Beach Park's Yacht Basin," *Tampa Tribune*, 11 January 1925.

40 "Oldsmar Well Equipped for Yachting Activities," *Tampa Tribune*, 11 January 1925.

41 "New Municipal Yacht Basin Proves Popular," *Tampa Tribune*, 13 February 1927.

42 E. C. Garvin to District Engineer, 2 May 1923, "Preliminary Examinations–St. Petersburg Harbor," box 3320, Records of the Office of the Chief of Engineers, Jacksonville District, RG 77, Atlanta, Georgia.

43 Arsenault, *St. Petersburg and the Florida Dream*, 199.

44 "Boating Memories Gain Steam," *Tampa Tribune*, 8 December 1984.

45 "Another Victory for 'Short Route,'" *St. Petersburg Times*, 7 January 1910.

46 "Want Bayshore Road to Manatee; Will Be Built," *Tampa Daily Times*, 31 July 1915.

47 Arsenault, *St. Petersburg and the Florida Dream*, 196–97. "30,000 Motor across Gandy Bridge Sunday," *St. Petersburg Times*, 24 November 1924. "$1,500,000 Is Being Spent to Enlarge Gandy Bridge," *St. Petersburg Times*, 12 September 1926.

48 Arsenault, *St. Petersburg and the Florida Dream*, 197–98.

49 "Snell Isle Improvements Being Rushed," *St. Petersburg Times*, 28 February 1926.

50 "Record Acreage Deal Brings Profit of $4,000,000 to Syndicate in Tampa," *Tampa Times*, 5 September 1925.

51 "$100,000,000 Improvement Planned for Tampa Beach," *Tampa Times*, 7 September 1925.

52 "Tampa Beach Brought Red Grange to Tampa," *Tampa Morning Tribune*, 9 February 1926.

53 "Tampa Beach Declared Bankrupt by Court Order," *Tampa Morning Tribune*, 26 July 1928.

54 "Bay Causeway Work Expected to Begin Feb. 1," *Tampa Morning Tribune*, 8 January 1929.

55 "Dredge Raises First Mile on Bay Causeway," *Tampa Sunday Tribune*, 14 April 1929.

56 Hampton Dunn, "Will Its Original Name Be Restored to 'Davis Causeway'?" *Sunland Tribune* 5, no. 7 (1979), https://scholarcommons.usf.edu/sunlandtribune/vol5/iss1/7.

57 "Bayshore Wall Project Gets Federal Okay," *Tampa Tribune*, 19 September 1935.

58 "Group Protests Proposed Beach along Bayshore," *Tampa Morning Tribune*, 22 May 1935.

59 "A Bayshore Nuisance," *Tampa Times*, 17 March 1945.

60 "Restore DeSoto Park Beach," *Tampa Times*, 13 June 1927.

61 "County to Develop Beach Created by Construction of 22nd Street Causeway," *Tampa Morning Tribune*, 8 June 1927. "County Sets Up Safeguards at Bathing Beach," *Tampa Morning Tribune*, 6 August 1927.

62 "Why Not Another Beach?" *Tampa Daily Times*, 10 June 1927. "Beach Proposed at Ballast Point," *Tampa Morning Tribune*, 29 June 1927. "Permission Denied for Bathing Beach," *Tampa Daily Times*, 2 August 1927.

63 "Bathing Beach Site," *Tampa Daily Times*, 12 April 1934.

64 "100 Socialites to Enter Tampa Beauty Contest," *Pittsburgh Courier*, 1 June 1935.

65 "Bathing Beach Hopes Delayed," *Tampa Times*, 27 January 1934.

66 Port Tampa Beach advertisement, *Tampa Daily Times*, 19 June 1926.

67 "Health Board Rejects Site for City Beach," *Tampa Morning Tribune*, 24 April 1934.

68 "Campbell Causeway Popular Spot for Picnickers," *St. Petersburg Times*, 28 August 1949.

69 "Rocky Point Owners Offer City Site for Tampa Beach," *Tampa Morning Tribune*, 4 September 1946. This offer eventually fell through, and the developers attempted to make their own public beach there.

70 "$1,000,000 Beach Plan for Tampa Is Approved," *Tampa Morning Tribune*, 17 February 1950.

71 "Frills in City Beach Plan Hit by Taxpayers' Group," *Tampa Morning Tribune*, 15 July 1950.

72 "Bill Drawn to Give City Right to Develop Beach," *Tampa Morning Tribune*, 24 April 1947. "Bay Oil Leases Said to Delay Tampa Beach," *Tampa Morning Tribune*, 27 December 1947. "22nd Street Causeway Is Proposed as Beach Site," *Tampa Morning Tribune*, 5 October 1946. "Water Tests Show 22nd St. Site Unsuitable for Beach," *Tampa Morning Tribune*, 3 April 1947.

73 "Tampa Bay Beach Plan Is Outlined," *Tampa Morning Tribune*, 19 March 1950.

74 "Frills in City Beach Plan Hit by Taxpayers' Group," *Tampa Morning Tribune*, 15 July 1950.

75 "Negro Request Snags Bathing Beach Plans," *Tampa Times*, 26 July 1950.

76 "Tampa Jaycees Pursue Variety of Local Topics," *Tampa Times*, 10 June 1955.

77 "Hearing Set on Building Waiver," *Tampa Tribune*, 25 January 1961.

78 "City Gets $6,156 Bid on Bathhouse," *Tampa Times*, 6 July 1961. The city honored the builder of the causeway, Ben T. Davis, by naming the beach for him in 1963.

79 "Beach Orderly," *Tampa Tribune*, 15 May 1961.

80 "Negro Beach Unsuited for Use; City Should Renovate and Make Fit," *St. Petersburg Times*, 26 March 1952.

81 "Negro Beach Site Plans Unchanged," *St. Petersburg Times*, 3 June 1955.

82 "City to Renovate Present South Mole Negro Beach; Drop Boca Ciega Plan," *St. Petersburg Times*, 8 July 1955.

83 Peyton L. Jones, "Struggle in the Sunshine City: The Movement for Racial Equality in St. Petersburg, Florida, 1955–1968" (MA thesis, University of South Florida, 2010), 9.

84 Jones, "Struggle in the Sunshine City," 10. "Kerosene-Soaked Cross Is Found," *St. Petersburg Times*, 9 June 1958.

85 "Rogero Says Work Starting Today on Negro Beach Near West End of Gandy Bridge," *Tampa Tribune*, 11 June 1958.

86 "State Building Negro Beach in Pinellas, Near Causeway," *Tampa Tribune*, 13 December 1947. "Court Action Planned to Halt Pinellas Negro Beach," *Tampa Tribune*, 20 December 1947.

87 "State and City Dedicate Negro Recreation Area along Campbell Parkway," *Tampa Tribune*, 13 June 1958. "Judge Denies Suit Dismissal by City," *St. Petersburg Times*, 1 September 1962.

88 Jones, "Struggle in the Sunshine City," 11–13.

89 Charles E. Dawson, *A Survey of the Tampa Bay Area*, Florida State Board of Conservation Technical Series no. 8 (Tallahassee: Florida State Board of Conservation, 1953), 9.

90 Bruce Stephenson, "'A Monstrous Desecration': Dredge and Fill in Boca Ciega Bay," in *Paradise Lost? The Environmental History of Florida*, ed. Jack E. Davis and Raymond Arsenault (Gainesville: University Press of Florida, 2005), 326–49, quotation on 342.

Chapter 5. Sewer

1 "Bigger is Better?" *60 Minutes*, 10 February 1974, https://www.youtube.com/watch?v
=PpW6AG3lyWY.

2 "Bigger is Better?" *60 Minutes*, 10 February 1974.

3 "Some Toes Will Bruise," *Tampa Tribune*, 15 October 1970.

4 "Bigger is Better?" *60 Minutes*, 10 February 1974.

5 "Gibbons Seeks U.S. Agency to Deodorize Bayshore," *Tampa Tribune*, 12 May 1966.
"Bay Seaweed Fumes Hit MacDill Guards," *Tampa Times*, 25 August 1961. "Rotting
Seaweed Creates Bayshore Stench, Staining Homes, Silverware, China," *Tampa Tri-
bune*, 19 August 1961.

6 "Bayshore Stench Won't Give Up in Easy Manner," *Tampa Tribune*, 11 March 1968.

7 Federal Water Pollution Control Administration, *Problems and Management of Water Quality in Hillsborough Bay, Florida* (Washington, D.C.: Government Printing
Office, 1969).

8 The earliest study that nailed down the problem in scientific terms was released in
1964. Florida State Board of Health, "A Study of the Causes of Obnoxious Odors:
Hillsborough Bay, Hillsborough County Florida," 1965, Sam Gibbons Papers, box
48, folder 9, USF Libraries, Special Collections, Tampa.

9 "Plague of Smells Mighty Real One," *Tampa Times*, 30 August 1915. "Gas Decayed
Fish Peels Paint from House Walls," *Tampa Tribune*, 30 August 1915.

10 "What Tampa Doesn't Need," *Tampa Weekly Tribune*, 4 August 1910. The Mallory
Wharf was located where the Tampa Convention Center is today.

11 "Tampa's New Sewerage System Plans Finished," *Tampa Morning Tribune*, 17 July
1913. "City's New Sewerage System Will Be One of the Best in the Country," *Tampa
Sunday Times*, 27 September 1914.

12 "Septic Tanks Put in Use," *St. Petersburg Times*, 28 December 1921.

13 "Health Board Urges Sewers on North Side," *St. Petersburg Times*, 16 June 1929.

14 "Levy Asserts City Sewage Pollutes Bay," *Tampa Morning Tribune*, 1 August 1927.

15 "Officials See Little Chance to Purify Bay," *Tampa Morning Tribune*, 8 May 1928.

16 "Gibsonton Board Wants Sewer Line Removed from Bay," *Tampa Morning Tribune*,
8 December 1928.

17 "Suit Charges Sewage Ruins Oyster Beds," *Tampa Morning Tribune*, 15 July 1931.

18 "R. E. Gibson Loses Suit against City," *Tampa Times*, 28 April 1934.

19 "Oysters Here Found in Dangerous State," *St. Petersburg Times*, 9 January 1925.

20 "Sewer Project Gains Support, Leaders Report," *Tampa Morning Tribune*, 25 September 1938.

21 "Oyster Eaters Are Warned of Typhoid," *Tampa Morning Tribune*, 22 November
1929.

22 "Health Officials Describe Drive on Polluted Oysters," *St. Petersburg Times*, 24
March 1940.

23 "Manatee River Clams, Oysters Condemned," *Tampa Morning Tribune*, 6 March
1942.

24 Charles E. Dawson, *A Survey of the Tampa Bay Area*, Florida State Board of Conservation Technical Series no. 8 (Tallahassee: Florida State Board of Conservation, 1953), 12, 22–23, 30–31. "Industry on the Bay Bottom," *Tampa Tribune*, 24 January 1965.

25 "Industry on the Bay Bottom," *Tampa Tribune*, 24 January 1965. J. Kneeland McNulty, William N. Lindall Jr., and James E. Sykes, "Cooperative Gulf of Mexico Estuarine Inventory and Study, Florida: Phase I, Area Description," National Oceanic and Atmospheric Administration (NOAA) Technical Report, NMFS circ. 368, November 1972, 87.

26 "Crisis Seen Over Interbay Sewage Line," *Tampa Morning Tribune*, 18 July 1951.

27 Alan J. Bliss, "Making a Sunbelt Place: Tampa, Florida, 1923–1964" (PhD. diss., University of Florida, 2010), 221–25.

28 "Bay Pollution Will Continue," *Tampa Times*, 21 June 1968.

29 Willie Crain to Jan Platt, 11 September 1986. Jan Platt Papers, box 27, folder 1, USF Libraries, Special Collections, Tampa.

30 "Women and Their Government: The City Sewage Disposal Plant," *St. Petersburg Times*, 12 January 1942.

31 "Polluted Waters of Spa Beach Called 'Public Health Hazard.'" *St. Petersburg Times*, 19 January 1949.

32 "Cans, Weeds, Oil Film Disgrace to Spa Beach," *St. Petersburg Times*, 12 February 1950.

33 "Sewage Plant Completion Set Nov. 1," *St. Petersburg Times*, 1 October 1954.

34 "MacDill Begins Operation of $416,000 Sewage Plant," *Tampa Tribune*, 3 February 1953.

35 "U.S. Report Cites Bay Pollution Crisis," *Tampa Tribune*, 21 January 1970.

36 "Pollution Board Performance Hit," *Tampa Tribune*, 14 January 1970.

37 "How Polluted Is Florida?" *Florida Trend*, December 1969, 18–35.

38 "Pollution Study Unit Meets," *Tampa Times*, 30 December 1970.

39 U.S. Army Corps of Engineers, "Draft Environmental Impact Statement: Oyster Shell Dredging, Tampa and Hillsborough Bays, Florida" (U.S. Army Corps of Engineers: Jacksonville, March 1974), 40.

40 "State's Coasts: A Dump," *Tampa Tribune*, 12 April 1970.

41 "Bay Pollution Is Noted at Gulfport, But in Small Quantity, Survey Shows," *St. Petersburg Times*, 22 September 1952.

42 "Chamber Group Finds Palm and Alafia Rivers Reeking with Industrial Waste," *Tampa Morning Tribune*, 1 August 1957.

43 "Rocky Creek Nice to Look at But Not To Swim," *Tampa Times*, 1 November 1968.

44 William McGuckin, *Biodegradable: Detergents and the Environment* (College Station: Texas A&M University Press, 1991), 18.

45 Federal Water Pollution Control Administration, *Problems and Management of Water Quality in Hillsborough Bay, Florida*, 3. See n. 7 above for full citation.

46 Arch Fredric Blakey, *The Florida Phosphate Industry: A History of the Development and Use of a Vital Mineral* (Cambridge: Harvard University Press, 1973), appendix c, part 2.

47 Anthony M. Opyrchal and Kung-Lee Wang, *Economic Significance of the Florida Phosphate Industry: An Input-Output Analysis*, U.S. Department of the Interior Information Circular 8850 (Washington, D.C.: Government Printing Office, 1981), 27.

48 My descriptions of these processes are taken from Florida Polytechnic University's Florida Industrial and Phosphate Research website, "Phosphate Primer," https://fipr.floridapoly.edu/about-us/phosphate-primer/index.php.

49 "Alafia River Property Owners Sue Company," *Tampa Daily Times*, 20 December 1952.

50 "Pollution Problem," *Tampa Sunday Tribune*, 18 July 1954.

51 "Alafia River Bill Clears Legislature," *Tampa Morning Tribune*, 3 June 1955.

52 John Cherry to Jan Platt, 6 July 1987, Jan Platt Papers, box 78, folder 2, USF Libraries, Special Collections, Tampa. This letter includes a packet of articles written by Cherry in 1970 about the phosphate industry. Cherry was an engineer working on phosphate pollution control and wanted to provide the articles to a journalist as a whistleblower. He decided not to send the letters at the time but released them to Platt and the editors of the *Tampa Tribune* and the *St. Petersburg Times* in 1987 in order to "get the truth out and get the costs of cleaning up Tampa's bays to be spread out to those who are responsible."

53 Federal Water Pollution Control Administration, *Problems and Management of Water Quality in Hillsborough Bay*, 3. See n. 7 above for full citation.

54 "Clogged River, Dead Lakes," *Tampa Times*, 5 May 1970.

55 Federal Water Pollution Control Administration, *Problems and Management of Water Quality in Hillsborough Bay*, 3. See n. 7 above for full citation.

56 "Mackerel in Hillsborough Bay," *Tampa Times*, 6 June 1969. See also, "Hefty Fish Seen in Tourney," *Tampa Times*, 6 June 1969.

57 Willie Crain to Jan Platt, 11 September 1986. Jan Platt Papers, box 27, folder 1, USF Libraries, Special Collections, Tampa.

58 John L. Taylor, John R. Hall, and Carl H. Saloman, "Mollusks and Benthic Environments in Hillsborough Bay, Florida," in U.S. Department of Commerce, *Fishery Bulletin* 68, no. 2 (March 1970): 191–202.

59 George S. Gandy, oral interview by Ann B. Hodgson, 2015, USF Libraries, Oral History Program (OHP), Tampa. In *The Tampa Bay Estuary: An Oral History of Community Collaboration to Restore Ecological Integrity*, audio file T43_00017, https://digitalcommons.usf.edu/tb_estuary_ohp/17.

60 James Albert Rodgers, oral interview by Ann B. Hodgson, 2015, USF Libraries, Oral History Program (OHP), Tampa. In *The Tampa Bay Estuary: An Oral History of Community Collaboration to Restore Ecological Integrity*, audio file T43_00010, https://digitalcommons.usf.edu/tb_estuary_ohp/10.

61 Thom Foley, "The Taming of the Hillsborough River: How Tampa Gained a Moat, Destroyed a Creek, and Forgot a River," *Tampa Bay History* 21, no. 1 (2007): 1–19.

62 "Record of Public Hearing, Preliminary Examination Report on Channels from Tampa Bay to Safety Harbor, Florida, and from Old Tampa Bay to Oldsmar, Florida," 1948, U.S. Army Corps of Engineers Records, box 62, folder J-827, National Archives, Atlanta, Georgia.

63 "Safety Harbor Chamber of Commerce to Army Corps of Engineers," 14 June 1950, Records of the Office of the Chief of Engineers, Jacksonville District, box 62, folder J-827, RG 77 National Archives, Atlanta, Georgia.

64 U.S. Department of the Interior, *Report of the Bureau of Commercial Fisheries Biological Station, St. Petersburg Beach, Florida: Fiscal Years, 1962–1964* (Washington, D.C.: Government Printing Office, 1966), 7.

65 U.S. Army Corps of Engineers, "Draft Environmental Impact Statement: Oyster Shell Dredging, Tampa and Hillsborough Bays, Florida" (U.S. Army Corps of Engineers: Jacksonville, March 1974), 45.

66 Robin R. Lewis III and Ernest D. Estevez, *The Ecology of Tampa Bay: An Estuarine Profile*, U.S. Fish and Wildlife Service Biological Report 85 (Washington, D.C.: Government Printing Office, 1988), 59.

67 Sandra L. Husar, "The West Indian Manatee (*Trichechus manatus*)" Wildlife Research Report 7 (Washington, D.C.: U.S. Department of the Interior, Fish and Wildlife Service, 1977). "Manatee Study Renewed in Crystal River Area," *St. Petersburg Times*, 11 December 1972.

68 Charles E. Dawson, *A Survey of the Tampa Bay Area*, 36. See n. 24 above for full citation.

69 J. Kneeland McNulty, William N. Lindall Jr., and James E. Sykes, "Cooperative Gulf of Mexico Estuarine Inventory and Study," 87. See n. 25 above for full citation.

Chapter 6. Cause

1 "Environmental Director Calls for Building Halt," *St. Petersburg Times*, 14 September 1973. "Pungent Pollution Picture Painted," *St. Petersburg Times*, 26 September 1973.

2 "Pollution Chief Backs Off When Jobs Are Threatened," *St. Petersburg Times*, 21 September 1973.

3 "Anti-Pollution Chief Defends Report," *St. Petersburg Times*, 28 September 1973.

4 "Pollution Director Fired; His Chief Aide Resigns," *St. Petersburg Times*, 7 March 1974.

5 "Pollution Chief Fought City Hall—And Lost," *St. Petersburg Times*, 8 March 1974.

6 Letter to the editor, *St. Petersburg Times*, 8 March 1974.

7 "Hicks Firing Angers Readers," *St. Petersburg Times*, 14 March 1974.

8 "Hicks Firing Angers Readers, *St. Petersburg Times*, 14 March 1974.

9 Elsie Ranck to editor, *St. Petersburg Times*, 30 March 1974.

10 "Tampa Officials Stunned by Abrupt Stewart Ouster," and "Shock Waves Roll over Stewart Firing," *Tampa Times*, 26 March 1974.

11 "Stewart: The People Loved Him," *Tampa Tribune*, 30 March 1974.

12 Samuel P. Hayes, *Beauty, Health, and Permanence: Environmental Politics in the United States, 1955–1985* (Cambridge: Cambridge University Press, 1987).

13 "Preserve the Birds," *St. Petersburg Times*, 14 September 1901.

14 T. Gilbert Pearson, "Uncle Sam's Birds," *American Museum Journal* 16, no. 6 (October 1916): 398. T. Gilbert Pearson, "Florida Bird Notes," *Bird Lore* 9, no. 1 (January–February 1907).

15 "Tarpon Key Converted into Government Bird Sanctuary," *St. Petersburg Times*, 14 August 1921. For Tippetts's work on behalf of conservation in St. Petersburg and the Bay Area, see Leslie Kemp Poole, "Katherine Bell Tippetts: A Female Voice for Conservation during Florida's Boom," *Tampa Bay History* 22, no. 1 (2008), https://digitalcommons.usf.edu/tampabayhistory/vol22/iss1/6. "Bird Key," *Forest and Stream* 66 (10 March 1906): 382.

16 H. R. Mills, "Report on the Tampa Bay Rookeries," *Florida Naturalist* 8, no. 1 (October 1934): 2.

17 "Chapter 4557: An Act to Regulate the Catching or Taking of Fish in the Waters of the State of Florida," *Acts and Resolutions Adopted by the Legislature of Florida at Its Sixth Regular Session under the Constitution of A.D. 1885* (Tallahassee: Florida Printing Company, 1897), 99–100.

18 "The County's Fish Interests," *Braidentown News*, 12 November 1897.

19 "The New Fish Law," *Pensacola Journal*, 20 July 1897.

20 "Fish Laws Violated," *Pensacola Journal*, 20 January 1905.

21 "Pelican Death Penalty Sought," *St. Petersburg Times*, 29 August 1926.

22 Donald Worster, *Nature's Economy: A History of Ecological Ideas*, 2nd ed. (Cambridge: Cambridge University Press, 1994).

23 Helen Rozwadowski, *Vast Expanses: A History of the Oceans* (London: Reaktion Books, 2018), 152, 188–213.

24 "Boating and Fishing in Florida," *Florida Trend* (July 1966), 29.

25 "Wylie's State 'Suicide' Story Evokes Comments," *St. Petersburg Times*, 6 January 1962.

26 Letter from Mary Ward, *Fairmount News*, 8 October 1959.

27 R. Bruce Stephenson, *Visions of Eden: Environmentalism, Urban Planning, and City Building in St. Petersburg, Florida, 1900–1995* (Columbus: Ohio State University Press, 1997), 134–35.

28 "Move Is Made to Unite All Anti-Fill Groups," *St. Petersburg Times*, 4 November 1955.

29 "Boca Ciega Fills Aired!" *St. Petersburg Times*, 28 September 1956.

30 "Maps of Fills To Be Shown at Open House," *St. Petersburg Times*, 5 October 1956.

31 R. Bruce Stephenson, *Visions of Eden*, 126–42, quotation on 127.

32 "Furen Fill Would Bring 'Disaster' to Bay Sea Life, Biologists Testify," *St. Petersburg Times*, 11 December 1956.

33 "Wylie's State 'Suicide' Story Evokes Comments," *St. Petersburg Times*, 6 January 1962.

34 "A Cry for Wilderness: Researcher Says Fills Do Grave Harm," *St. Petersburg Times*, 25 January 1965.

35 "New Anti-Fill Group Starts Campaign," *St. Petersburg Times*, 9 January 1965.

36 "On the Waterfront," *St. Petersburg Times*, 23 January 1965.

37 "Strong Arguments Exist to Support Bay-to-Lake Conversion Plan," *Tampa Times*, 16 February 1957.

38 "Upper Bay Lake Project a Legislative Must," *Tampa Times*, 3 April 1957.

39 Lake Seminole Fact Sheet, https://www.pinellascounty.org/environmental/pdf/lake -seminole-fact-sheet.pdf.

40 "Volume of Fresh Water May Decide Bay Lake Plan," *Tampa Tribune*, 6 February 1958.

41 "Fresh Water Lake Plan Okayed by District Board," *Tampa Tribune*, 18 February 1965.

42 "Oldsmar, Safety Harbor Mayors Say Residents Oppose Creation of Lake," *Tampa Tribune*, 10 April 1958.

43 "Upper Bay Conversion to Lake Draws Objections," *Tampa Tribune*, 11 April 1965.

44 "Foes of Bay Lake Organize," *Tampa Tribune*, 28 April 1965.

45 "Foes of Bay Lake Wire Officials," *Tampa Tribune*, 29 April 1965.

46 G. M. Kindred to District Engineer, 27 May 1965, Sam Gibbons Papers, box 10, folder 6, USF Libraries, Special Collections, Tampa.

47 Undated notes, Sam Gibbons Papers, box 10, folder 6, USF Libraries, Special Collections, Tampa.

48 George K. Reid, *Ecology of Inland Waters and Estuaries* (New York: Reinhold, 1961). "Dr. Reid Is Educator, Author, and Outdoorsman," *St. Petersburg Times*, 25 January 1965.

49 "Presbyterian College Offers Marine Ecology," *St. Petersburg Times*, 11 November 1963.

50 Sylvia Earle Biography, National Women's History Museum, https://www .womenshistory.org/education-resources/biographies/sylvia-earle.

51 "'Country Boy' Asks County to Leave Tampa Bay As Is," *Tampa Times*, 23 June 1965. Taft and his students later drilled core samples north of the Courtney Campbell Causeway to prove the problems. They found two features that proved him right: salt deposits measuring six feet in depth under the bay bottom that would continue to leech into the water and an even deeper layer of impermeable clay that would prevent water from moving into the aquifer at all. "Bay Lake 'Wouldn't Get Fresh,'" *St. Petersburg Times*, 27 April 1966.

52 "Upper Bay Conversion to Lake Draws Objections," *Tampa Tribune*, 11 April 1965.

53 "Bay 'Lake' Sparks Big Problems," *St. Petersburg Times*, 5 April 1965.

54 "Upper Bay Conversion to Lake Draws Objections," *Tampa Tribune*, 11 April 1965.

55 "'Country Boy' Asks County to Leave Tampa Bay As Is," *Tampa Times*, 23 June 1965.

56 "Who Is the Real Fear Monger?" *Tampa Times*, 28 April 1966.

57 WTVT editorial, "Bay Lake Issue," 28 May 1965, Sam Gibbons Papers, box 10, folder 6, USF Libraries, Special Collections, Tampa.

58 "Bay Lake Critics Rapped," *St. Petersburg Times*, 13 May 1965.

59 "Pinellas Takes Definite Stand Opposing Bay Project," *Tampa Times*, 16 November 1965.

60 "Hodges Opposes Upper Tampa Bay Fresh Water Lake Project," *Tampa Tribune*, 5 August 1966.

61 "Steps Taken to Abandon Tampa Bay-Lake Project," *Tampa Tribune*, 1 March 1968.

62 James B. Lackey, Clark I. Cross, and Ronald G. Schmidt, "Socioeconomic Impact of Creating Tampa Bay Fresh Water Lake," Report for the Southwest Florida Wa-

ter Management District by Reynolds, Smith, and Hill, December 1968, Tampa-Hillsborough County Public Library, Special Collections.

63 "Developer Tells Plans for 12,000 Homes Here," *Tampa Tribune*, 15 May 1959.

64 "$1 Billion Seen in Bower Plan," *Tampa Tribune*, 20 September 1969. "Bayport Colony: Yes! Well-rounded Community, Healthier Ecology Pledged," *Tampa Times*, 3 March 1970.

65 For the history of land development in this era, see Jason Vuic, *The Swamp Peddlers: How Lot Sellers, Land Scammers, and Retirees Built Modern Florida and Transformed the American Dream* (Chapel Hill: University of North Carolina Press, 2021).

66 "Apollo Beach: Poised to Grow," *St. Petersburg Times*, 8 March 1987.

67 "Dana Shores Ready to Blossom into Community," *Tampa Times*, 12 October 1957.

68 "Huge Dredge Fill Sought," *Tampa Times*, 12 September 1969.

69 "Bower Project Draws Warning," *Tampa Times*, 16 December 1969.

70 "Big Project Seen as Death Threat to Bay," *Tampa Tribune*, 23 January 1970.

71 "Bower Fill Plan Hit," *Tampa Tribune*, 7 April 1970.

72 "Ecologists Rush to Shelter Sharp-Tongued Prof. Taft," *Tampa Times*, 9 March 1971.

73 "Bay Port Project Draws Opposition," *Tampa Tribune*, 22 March 1970.

74 "Open Letter to Our Citizens," *Tampa Tribune*, 17 April 1970.

75 "Bay Port Dredge Project Comes Under New Attacks," *Tampa Tribune*, 1 March 1970.

76 "Students to Hold Crusade," *St. Petersburg Times*, 8 April 1970.

77 "Save Our Bay: Residents Fight to Prevent Pollution of Old Tampa Bay," *Tampa Tribune*, 19 April 1970.

78 "Save Our Bay: Residents Fight to Prevent Pollution of Old Tampa Bay," *Tampa Tribune*, 19 April 1970.

79 "Bay Savers: The Politics of Pollution," *Tampa Times*, 8 June 1971.

80 "Bay Savers: The Politics of Pollution," *Tampa Times*, 8 June 1971.

81 "A 'Killer' Exposed," *Tampa Times*, 11 November 1969.

82 "Save Our Bay: Residents Fight to Prevent Pollution of Old Tampa Bay," *Tampa Tribune*, 19 April 1970.

83 "300 Boo, Catcall at Pollution Meeting," *Tampa Times*, 26 February 1970.

84 "New SOB Group Incorporates," *Tampa Tribune*, 24 June 1970. "Bay Savers: The Politics of Pollution," *Tampa Times*, 8 June 1971.

85 "Lawsuit Challenges Bay Port Rezoning," *Tampa Tribune*, 28 February 1970. "Bay Port Decision Delayed but Rezoning Appears Likely," *Tampa Tribune*, 5 March 1970. "Showdown Looms on Old Tampa Bay Fill Project Today," *St. Petersburg Times*, 17 April 1970. "Bay Port Challenged on Finances," *Tampa Tribune*, 18 April 1970.

86 "Bay Port Idea Still Live Issue," *Tampa Tribune*, 30 January 1971.

87 "Save Our Bay Sues Commission," *Tampa Times*, 7 September 1971.

88 "Restraining Order Expected on Tract," *St. Petersburg Times*, 22 September 1971.

89 "Hillsborough Fill Removal: A Precedent?" *St. Petersburg Times*, 20 October 1972.

90 "Governor, Environmentalists Hail Purchase of Tract on Tampa Bay," *St. Petersburg Times*, 11 December 1985.

Chapter 7. Estuary

1 "Tanker's Oil Stains Tampa Bay," *Tampa Tribune*, 14 February 1970.

2 "Disastrous Oil Spill of 1970 Made National Headlines," *St. Petersburg Times*, 2 April 2003.

3 "The Dirty Dilemmas of Oil Spills," *Life*, 6 March 1970, 28–35.

4 "Oil Spill . . . $250 Million Suit Filed," *Tampa Tribune*, 14 March 1970. "Oil Spill Suits Crowd Court Here," *Tampa Tribune*, 26 March 1970.

5 "Investigation by State Set after Spill Kills Fish," *Tampa Times*, 24 February 1970.

6 "Bay Oil Slick Has a Silver Lining," *Tampa Tribune*, 22 February 1970.

7 "Pollution Safeguards up to Officials," *Tampa Tribune*, 17 September 1972.

8 Roy H. Lasris, ed., *Florida Water Resources Law: A Bibliography* (Washington, D.C.: Government Printing Office, 1977), 28–30.

9 "Tampa Bay Area Study Group Project," USF Libraries, Special Collections, Tampa, https://digitalcommons.usf.edu/basgp/.

10 "Multitudes Came to Hear of Pollution," *Tampa Tribune*, 27 February 1970.

11 "Pollution Board Performance Hit," *Tampa Tribune*, 14 January 1970.

12 "Bay Planners to Oppose Sewage Treatment Act," *St. Petersburg Times*, 22 July 1972.

13 Richard D. Garrity, oral interview by Ann B. Hodgson, 2015, USF Libraries, Oral History Program (OHP), Tampa. In *The Tampa Bay Estuary: An Oral History of Community Collaboration to Restore Ecological Integrity*, audio file T43–00013, https://digitalcommons.usf.edu/tb_estuary_ohp/13.

14 Howard L. Rhodes, "Municipal Treatment Requirements and Practices to Maintain Water Quality in the Tampa and Escambia Bay Areas," in U.S. Environmental Protection Agency, *Proceedings of the Seventh United States/Japan Conference on Sewage Treatment Technology, May 19–21, 1980* (Washington, D.C.: Government Printing Office, 1980), 443.

15 "Ecology Coalition Organized," *Tampa Times*, 9 May 1973.

16 George S. Gandy, oral interview by Ann B. Hodgson, 2015, USF Libraries, Oral History Program (OHP), Tampa. In *The Tampa Bay Estuary: An Oral History of Community Collaboration to Restore Ecological Integrity*, audio file T43_00017, https://digitalcommons.usf.edu/tb_estuary_ohp/17.

17 "Deepening of Channels Poses Ordeal for Tampa Bay," *St. Petersburg Times*, 15 March 1974.

18 "Harbor-Dredging Plans Criticized," *Tampa Tribune*, 17 April 1974. "Panel Will Advise on Harbor Project," *St. Petersburg Times*, 22 April 1974.

19 WDAE editorial transcript, Sam Gibbons Papers, box 92, folder 8, USF Libraries, Special Collections, Tampa.

20 "Harbor-Dredging Plans Criticized," *Tampa Tribune*, 17 April 1974. "Panel Will Advise on Harbor Project," *St. Petersburg Times*, 22 April 1974. "Joseph Simon, Scientist and USF Biology Pioneer," *St. Petersburg Times*, 24 March 2004.

21 "Panel Will Advise on Harbor Project," *St. Petersburg Times*, 22 April 1974.

22 Robin Lewis to Hal Cusick, 6 February 1975, Sam Gibbons Papers, box 92, folder 10, USF Libraries, Special Collections, Tampa.

23 "Photos Bring Environmentalists, Business Together," *Tampa Tribune*, 12 December 1974.

24 "Port Project Suit Feared," *Tampa Tribune*, 4 January 1975.

25 "Harbor Battle Resolved," *Tampa Tribune*, 11 March 1975.

26 "House Okays Pinellas Aquatic Preserve Bill," *St. Petersburg Times*, 26 May 1970.

27 "Suddenly, Almost Everybody's Anti-Pollution," *Tampa Tribune*, 2 April 1967.

28 "Spicola Cited for His Interest in Conservation," *Tampa Times*, 30 September 1974.

29 Sigrid Tidmore, "Jan Platt: The Young Girl and the Sea," *Bay Soundings*, https://baysoundings.com/jan-platt-the-young-girl-and-the-sea/.

30 Jan Platt political advertisement, *Tampa Times*, 25 September 1978.

31 "Jan Platt Obviously the One," *Tampa Times*, 21 August 1978.

32 Proposed map of Old Tampa Bay Regional Park, 1969, Jan Platt Papers, box 108, folder 7, USF Libraries, Special Collections, Tampa.

33 "New Parks to Emphasize Environment," *Tampa Tribune*, 1 January 1979.

34 Anita Scharf to Bob Curry, 23 June 1979, Jan Platt Papers, box 108, folder 7, USF Libraries, Special Collections, Tampa.

35 John B. Sharpe to Hillsborough County Department of Parks and Recreation, 2 July 1979, Jan Platt Papers, box 10, folder 1, USF Libraries, Special Collections, Tampa.

36 "Weedon Island," *St. Petersburg Times*, 22 November 1974. "Nature's Pageant Envelops Weedon Island," *St. Petersburg Times*, 22 January 1978. "Weedon Island Opening to Public after a Year of Cleanup," *St. Petersburg Times*, 28 December 1980.

37 "Land Use: Voters Okay Purchase of Fragile Land," *St. Petersburg Times*, 5 November 1980.

38 "Fish Kill Worsens in Old Tampa Bay," *St. Petersburg Times*, 23 September 1977.

39 "Dead Sea," *St. Petersburg Times*, 7 April 1976.

40 "Tampa's Fishing Fleet's End Nears," *Tampa Times*, 3 June 1969.

41 Robert H. Boyle and Rose Mary Mechem, "There's Trouble in Paradise," *Sports Illustrated*, 9 February 1981.

42 Sara-Ann F. Treat, Joseph L. Simon, Roy R. Lewis III, and Robert L. Whitman Jr., eds., *Proceedings–Tampa Bay Area Scientific Information Symposium, May 1982*, Florida Sea Grant College report 65 (n.p.: Bellwether Press, 1985), ii.

43 "Platt Says Comprehensive Study of Bay 'Overdue,'" *Tampa Tribune*, 9 May 1982.

44 "Commission's Plan Could Bring Back the Gleam," *Tampa Tribune*, 16 May 1985.

45 Tampa Bay Management Study Commission, "The Future of Tampa Bay," report to the Florida Legislature and the Tampa Bay Regional Planning Council, 1985, Jan Platt Papers, box 33, folder 1, USF Libraries, Special Collections, Tampa.

46 Peter A. Clark, oral interview by Ann B. Hodgson, USF Libraries, Oral History Program (OHP), Tampa. In *The Tampa Bay Estuary: An Oral History of Community Collaboration to Restore Ecological Integrity*, audio file T43_0008, https://digitalcommons.usf.edu/tb_estuary_ohp/8.

47 Florida Department of Natural Resources, "Cockroach Bay Aquatic Preserve Management Plan, April 1987," ii, Jan Platt Papers, box 80, folder 6, USF Libraries, Special Collections, Tampa.

48 "Shell Pit Developer Denied Permit to Dredge, Fill Aquatic Preserve," *Tampa Tribune*, 9 January 1986.
49 "Developer Puts Forth Marina Plan Again," *Tampa Tribune*, 1 September 1986.
50 "State Says No to Marina near Little Cockroach Bay," *St. Petersburg Times*, 17 June 1987.
51 "Dreams Can Enrich the Community," *Tampa Tribune*, 16 April 1986. "Mangrove Bay Marina Needed," *Tampa Tribune*, 21 May 1986.
52 Calendar of Events, *Tampa Tribune*, 19 May 1985.
53 "Fisherman Fishes for Sun City Supporters," *Sun City Shopper Observer News*, 10 September 1986.
54 "Gus Muench Worships Where Others Prey," *St. Petersburg Times*, 14 June 1987.
55 Sierra Club Tampa Bay Group Newsletter, May 1986, Jan Platt Papers, box 78, folder 2, USF Libraries, Special Collections, Tampa.
56 Hillsborough County Board of County Commissioners Agenda Item: Conservation and Recreation Land Program—Cockroach Bay Islands, 11 July 1986, Jan Platt Papers, box 78, folder 1, USF Libraries, Special Collections, Tampa. "Public Lands Program History," Florida Division of Historical Resources, https://dos.myflorida.com/historical/archaeology/public-lands/program-history/.
57 Gus Muench to Greg Brock, 13 July 1986, Jan Platt Papers, box 78, folder 2, USF Libraries, Special Collections, Tampa.
58 "Let's Try Again to Buy Singular Cockroach Key," *Tampa Tribune*, 13 June 1986.
59 Rodney Colson to Greg Brock, 23 July 1986, Jan Platt Papers, box 78, folder 1, USF Libraries, Special Collections, Tampa.
60 Memo to Board of County Commissioners, 17 August 1989, Jan Platt Papers, box 83, folder 1, USF Libraries, Special Collections, Tampa.
61 "TECO Picks Bay Site for Plant," *St. Petersburg Times*, 22 December 1988.
62 "Power Plant Plan Raises Concerns for Environment," *Tampa Tribune*, 22 December 1988.
63 ManaSota-88 Newsletter, January 1989, Jan Platt Papers, box 80, folder 6, USF Libraries, Special Collections, Tampa.
64 "Agency Says Plan for 3rd TECO Plant Bad for Tampa Bay," *Tampa Tribune*, 27 October 1988.
65 Jan Platt to Pam Iorio, 24 October 1988, Jan Platt Papers, box 80, folder 6, USF Libraries, Special Collections, Tampa.
66 "Regional County Fights County on Plant Site," *Tampa Tribune*, 14 August 1990.
67 Richard M. Eckenrod, oral interview by Ann B. Hodgson, 2015, USF Libraries, Oral History Program (OHP), Tampa. In *The Tampa Bay Estuary: An Oral History of Community Collaboration to Restore Ecological Integrity*, audio file T43_00012, https://digitalcommons.usf.edu/tb_estuary_ohp/12.
68 Richard M. Eckenrod, oral interview by Ann B. Hodgson, 2015, USF Libraries, Oral History Program (OHP), Tampa, https://digitalcommons.usf.edu/tb_estuary_ohp/12.
69 "Factory Owner Fined for '04 Spill," *St. Petersburg Times*, 2 July 2005.

70 Peter A. Clark, oral interview by Ann B. Hodgson, USF Libraries, Oral History Program (OHP), Tampa. In *The Tampa Bay Estuary: An Oral History of Community Collaboration to Restore Ecological Integrity*, audio file T43_0008, https://digitalcommons.usf.edu/tb_estuary_ohp/8.

Epilogue

1 This account comes largely from Ellen Bausback, "A Timeline of the Piney Point Wastewater Disaster," 22 April 2022, Florida Museum of Natural History, https://www.floridamuseum.ufl.edu/earth-systems/blog/a-timeline-of-the-piney-point-wastewater-disaster/.

2 "Phosphate Plant Leak to be Funneled to Bay," *Tampa Bay Times*, 31 March 2021. "Tampa Bay Bucks Global Trend toward Ocean Acidification," *Bay Soundings*, undated, https://baysoundings.com/tampa-bay-bucks-global-trend-toward-ocean-acidification/.

3 "Months after Piney Point Wastewater Disaster, a Massive Red Tide Has Killed 1,600-plus Tons of Marine Life—A Huge Blow to the Local Fishing Community," *The Counter*, 29 July 2021, https://thecounter.org/piney-point-wastewater-disaster-red-tide-dead-florida-marine-life-fishermen-shrimpers/.

4 "Tampa Bay among Fastest-Growing Large U.S. Metro Areas," *Business Observer*, 23 December 2022, https://www.businessobserverfl.com/news/2022/dec/23/tampa-bay-among-fastest-growing-large-u-s-metro-areas/. This measurement includes counties beyond the immediate Bay Area.

5 "Tampa Bay Lost 12% of Its Seagrass in 2 years; Some Areas at Historic Low, Study Shows," *Tampa Bay Times*, 13 February 2023.

6 "Tampa Bay's Coming Storm," *Washington Post*, 28 July 2017.

INDEX

Evan P. Bennett is associate professor of history at Florida Atlantic University in Boca Raton. He is the author of *When Tobacco Was King: Families, Farm Labor, and Federal Policy in the Piedmont* and coeditor of *Beyond Forty Acres and a Mule: African American Landowning Families since Reconstruction.*

Florida in Focus

Edited by Andrew K. Frank

Books in this series provide original and lively introductions to a range of topics in Florida history. Written by established scholars and using original research, the books draw upon current scholarly developments to situate subjects in a broad historical context.

Before the Pioneers: Indians, Settlers, Slaves, and the Founding of Miami, by Andrew K. Frank (2017)
Millard Fillmore Caldwell: Governing on the Wrong Side of History, by Gary R. Mormino (2020)
Tampa Bay: The Story of an Estuary and Its People, by Evan Bennett (2024)